£12.99

AQA
GCSE BIOLOGY

Vic Pruden, Jennifer Burnett,
Judy Crane, Christine Woodward

Hodder & Stoughton

30130504249714

D0317395

570
PRU

Photo acknowledgements

The publishers would like to thank the following individuals, institutions and companies for permission to reproduce photographs in this book. Every effort has been made to trace ownership of copyright. The publishers would be happy to make arrangements with any copyright holder whom it has not been possible to contact:

Action Plus (34 top, 132 top); Australian Museum, Nature Focus (136 top, 138); Biophoto Associates (148 bottom right); Bruce Coleman Ltd (102 all, 112 fennec fox, 120 middle, 127 bottom left, 128 top left, 140, 141, 148 bottom left, 159); Corbis (84); Hodder & Stoughton (100); Holt Studios (73 both, 74 both, 75, 83, 96, 98 bottom); Life File (103 right, 120 right, 127 bottom right, 158, 166, 170 bottom, 187 top, 190); Museum of London (60); Natural History Museum (104 top,); Oxford Scientific Films (137, 139 left, 155 bottom, 156 top, 172 top); Panos Pictures (193); R D Battersby (189 bottom); RSPCA Photo Library (98 top left); Science Photo Library (1, 3, 11, 19, 23, 31 both, 34 bottom, 43 both, 57, 58, 59, 76, 82, 86, 92, 94, 95, 97, 98 top right, 103 left, 104 middle, 112 arctic fox, cacti, 113 both, 114, 116, 120 left, 130, 132 bottom left & right, 134 middle left & right, 139 right, 150 all, 152 all, 153 all, 155 top, 157 top right, 164, 167, 168, 169 all, 170 top left & right, 171, 172 bottom, 174, 175, 176 both, 178, 179, 187 bottom, 189 top, 192, 196); Skulls Unlimited International Inc. (156 bottom, 157 top left & top middle); Steve Zinsky (128 right); Wellcome Trust (134 top left, middle & right, 165, 173).

Orders: please contact Bookpoint Ltd, 130 Milton Park, Abingdon, Oxon OX14 4SB.
Telephone: (44) 01235 827720. Fax: (44) 01235 400454. Lines are open from 9.00–6.00, Monday to Saturday, with a 24 hour message answering service.

British Library Cataloguing in Publication Data
A catalogue record for this title is available from the British Library

ISBN 0 340 84783 2

First published 2002
Impression number   10   9   8   7   6   5   4   3   2
Year   2008   2007   2006   2005   2004   2003   2002

Cover illustration by Sarah Jones at Debut Art
Typeset by J&L Composition Ltd, Filey, North Yorkshire.
Printed in Italy for Hodder & Stoughton Educational, a division of Hodder Headline, 338 Euston Road, London NW1 3BH.

Contents

About this book

The contents

The contents of this book are designed to cover all aspects of the knowledge and understanding required by the AQA GCSE specifications in Biology (Co-ordinated) and Biology (Modular).

The subject content required by the KS4 Double Award specification for Life and living processes attainment target is produced in a format identical to that used in the Hodder and Stoughton textbook *AQA GCSE Science*. This core material is supplemented by the additional subject content required for the specification in GCSE Biology.

What is in each chapter?

At the beginning of each chapter is a list of **Key Terms**. Where used for the first time, these key terms are emboldened. Some of the key terms are coloured. These are the extra terms you will need to know if you are going to be entered for the Higher tier papers in the final examination. All the key terms together with their meanings are also found in the **Glossary** on pages 201–209.

The contents of each chapter are divided into several sections. Each section concentrates on one topic. A symbol at the start of each section shows clearly which topic from the co-ordinated and modular courses is being targeted.

You will see a number of **Did you know?** boxes throughout each chapter. You will not have to learn the information in these boxes, but they are there to give extra interest to the topic.

At the end of various sections, you will find a number of **Topic Questions**. Because the topic questions have been designed to produce answers that you could use as a set of revision notes, it is recommended that you write down the questions as well as the answers. The questions written on a yellow background are the more demanding questions, expected to be answered if you are a grade B/A/A★ student. Don't worry if you have to re-read the topic again when you try to answer these questions. This will help you to learn the work.

At the end of each chapter is a **Summary**. The summary provides a brief analysis of the important points covered in the section.

Completing each chapter are some **GCSE questions** taken from past AQA (SEG) or past AQA (NEAB) examination papers. The questions written on a yellow background are the more demanding questions expected to be answered if you are a grade B/A/A★ student. Answering the GCSE questions will help give you an idea of what is wanted when you take your final science examination. Again, do not worry if you have to go back to read the work again. The examination questions may well test you on knowledge not included in the particular chapter. Don't worry – look through the other chapters to find the extra information you need to complete your answer.

Specification Matching Grid

Table 1.1   Life and Living Processes

Chapter		Section		AQA specification references	
			Content	**Co-ordinated**	**Modular**
1	Cell activity	1.1	Animal cells	10.1	01 (10.1)
		1.2	Plant cells	10.1	02 (11.1)
		1.3	Diffusion	10.2	01 (10.6)
		1.4	Osmosis and active transport	10.2	02 (11.3)
		1.5	Cell division	10.3	04 (13.1)
2	Humans as	2.1	Nutrition	10.4	01 (10.2)
	organisms	2.2	Composition of the blood	10.5	01 (10.4)
		2.3	The circulatory system	10.5	01 (10.4)
		2.4	Breathing	10.6	01 (10.3)
		2.5	Respiration	10.7	01 (10.3)
3	Response,	3.1	Nervous system	10.8	02 (11.5)
	co-ordination and	3.2	Hormones and the menstrual cycle	10.9	02/04 (11.6, 13.5)
	health	3.3	Hormones and diabetes	10.9	02 (11.6)
		3.4	Homeostasis	10.10	02 (11.6)
		3.5	Fighting disease	10.11	01 (10.5)
		3.6	Drugs	10.12	02 (11.7)
4	Green plants as	4.1	Plant nutrition	10.13	02 (11.2)
	organisms	4.2	Plant hormones	10.14	02 (11.4)
		4.3	Transport and water relations	10.15	02 (11.3)
5	Variation,	5.1	Variation	10.16	04 (13.1)
	inheritance and	5.2	Genes	10.17	04 (13.1)
	evolution	5.3	DNA	10.17	04 (13.4)
		5.4	Controlling inheritance	10.18	04 (13.2)
		5.5	Evolution	10.19	04 (13.3)
6	Living things in	6.1	Adaptation and competition	10.20	03 (12.1)
	their environment	6.2	Human impact on the environment	10.21	03 (12.4)
		6.3	Energy and nutrient transfer	10.22	03 (12.2)
		6.4	Nutrient cycles	10.23	03 (12.3)

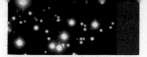

Specification Matching Grid

Chapter		Section	Content	AQA specification references	
				Co-ordinated	Modular
*7	Locomotion	7.1	How vertebrates are adapted for movement	10.24	19 (14.1)
		7.2	How exercise can benefit health	10.24	19 (14.2)
		7.3	How fish are adapted for swimming	10.25	19 (14.3)
		7.4	How birds are adapted for flying	10.25	19 (14.4)
*8	Feeding	8.1	How some invertebrates are adapted for feeding	10.26	19 (14.5)
		8.2	How some mammals are adapted for feeding	10.27	19 (14.6)
*9	Controlling disease	9.1	How biology has helped us to control infectious diseases	10.28	20 (15.1)
		9.2	How biology has helped us to treat kidney disease	10.29	20 (15.2)
*10	Using microorganisms	10.1	How microorganisms are used to make food and drink	10.30	20 (15.3)
		10.2	Other useful substances made using microorganisms	10.31	20 (15.4)
		10.3	Using microorganisms safely	10.30	20 (15.5)

*Additional content which extends the National Curriculum core Biology, Sc2, Life Processes and Living Things

Ideas and evidence in Science

You will find that many sections contain information which is marked with a bell and a vertical stripe in the margin. This is material to support the 'Ideas and Evidence in Science' part of your course. It will provide you with information about:

- how scientific ideas were developed and presented,
- how scientific arguments can arise from different ways of interpreting the evidence,
- ways in which scientific ideas may be affected by the contexts in which it takes place (for example, social, historical, moral and spiritual) and how these contexts may affect whether or not ideas are accepted,
- the problems science has in dealing with industrial, social and environmental questions, including the kinds of questions science can and cannot answer, uncertainties in scientific knowledge, and the ethical issues involved.

Each of the 'Ideas and evidence' contexts needed for whatever course you are following is included in this book. A guide to these contexts, is given in the table below

Table 1.4   Contexts for the delivery of 'Ideas and evidence' in Life and Living Processes

Section	DA	Core/HT	Context
3.2	✓	core	Benefits and problems caused by the use of hormones to control fertility
3.5	✓	core	How living conditions and life style are related to the spread of disease
3.6	✓	core	Why the link between smoking tobacco and lung cancer gradually became accepted
5.1	✓	core	Why Mendel proposed the idea of separately inherited factors and why this discovery was not immediately recognised
5.4	✓	core	The economic, ethical and social issues raised by the development of cloning and genetic engineering
5.5	✓	core	How fossil evidence supports the theory of evolution
5.5	✓	core	How over-use of antibiotics can lead to the evolution of resistant bacteria
5.5	✓	core	Why Darwin's theory of natural selection was only gradually accepted
6.2	✓	core	How the managing of food production for human needs is a compromise between competing priorities
6.2	✓	core	Some of the major environmental issues facing society
6.3	✓	core	Problems involved with the large scale production of food

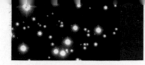

Ideas and evidence in Science

Section	DA	Core/ HT	Context
9.1	✓	core	How Pasteur showed that decay and disease are caused by living organisms
9.1	✓	core	How the treatment of disease has changed as a result of increased understanding of the action of antibiotics and immunity
9.1	✓	core	The advantages and disadvantages of being vaccinated against a particular disease
9.2	✓	core	The advantages and disadvantages of treating kidney failure by dialysis or kidney transplants
10.2	✓	core	The economic and environmental advantages and disadvantages of the production of fuels by fermentation and their uses

Some hints about doing well in the final written examinations

Some frequently used command words and what they mean

Before you can answer a question, you need to know what is expected. Question-writers use command words or phrases that inform you of the style of answer they expect you to give. A list of the most frequently used command words and phrases is given below. Question-writers assume that you have learned the meanings of the words or phrases.

Calculate or **work out** .... means that a calculation is needed together with a numerical answer.

Compare .... means that a description is needed of the similarities and/or differences in the information that has been provided.

Complete .... means that spaces in a diagram, a table or a sentence or sentences need to be completed.

Describe .... means that the important points about the particular topic must be provided.

Draw a bar chart .... – if the axes are already labelled and scales have been given then the values given must be plotted as bars.
– if the axes are labelled but no scales have been given then scales need to be added and the values given need to be plotted as bars.

Draw a graph .... – if the axes are already labelled and scales have been given then the values given need to be plotted as points and a line (or curve) appropriate to the points plotted must be drawn.
– if the axes are labelled but no scales have been given then the scales need to be added, the values given must be plotted as points and a line (or curve) appropriate to the points plotted must be drawn.

Explain how or **Explain why** .... means that scientific theory must be used to show an understanding of how or why something happens.

Give a reason or **How** or **Why** ... means that the answer requires a cause for something happening based on scientific theory.

Give or **Name** or **State** or .... means that a short answer with no supporting **Write down** scientific theory is needed.

List .... means that a number of short answers are needed, each one being written one after the other.

Predict or **Suggest** .... means that the answer is based on a *consideration* of various pieces of information and suggesting, without supporting theory, what is likely to happen.

Sketch a graph .... means that a line (or curve) is to be drawn to show a trend or pattern without the need to plot a series of points.

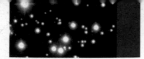

Exam hints

Use the information .... means that the answer must be based on the information provided in the question.

Use your understanding of ... to this is the science topic around which the answer needs to be built.

What is meant by .... means that the answer is likely to be a definition.

Some more hints

Obviously if you want to do well you need to have learned and understood as much as you can. However here are some hints about answering questions.

- Do not rush – no marks are awarded for finishing first. A paper worth 100 marks is designed to allow you about 90 minutes to finish it. This means that you have nearly one minute of time to think and write down 1-marks worth of answer.

- Read each question carefully at least twice before you write down your answer. If you need to do rough working to sort out your thoughts use the gaps in the margins – but make sure you put a line through this rough working.

- Look at the number of marks awarded for each part of each question. Each mark is given for a different piece of information:
  - *1 mark* means that one piece of information is needed.
  - *2 marks* mean that two pieces of information are needed etc.

- Lots of questions ask you to give a reason for something or to explain something. Such answers are usually worth 2 or more marks. Your answers to these should include a 'because' part.

- Do not throw away marks. Marks are often given for:
  - units such as joules, °C, ohms etc. Learn all the units and what they measure.
  - the names and symbols of chemical elements – so learn them
  - equations, such as 'potential difference = current × resistance' – so learn the equations and how to use them in calculations. Remember that all equations need an '=' sign in them.

- If you are writing an answer that needs several sentences, make sure that each sentence is saying something new and is not just rewording an earlier sentence.

- Try to avoid using the words 'it', 'they' or 'them' in an answer. The marker may find it difficult to understand what you mean.

- Take care when you are drawing graphs. Make sure all the points are correctly plotted. When you draw in the line for your points use a pencil with a fine point and try to draw the complete line in one go.

- If you have learned the work you should finish the paper in good time. Go through the paper again and check what you have written – it could save you throwing away some marks for silly mistakes.

Chapter 1
Cell activity

aerobic respiration · active transport · allele · alveolus · asexual reproduction · cell · cellulose · cell membrane · cell wall · chlorophyll · chloroplast · chromosome · concentration gradient · cytoplasm · diffusion · enzyme · fertilisation · gamete · gene · meiosis · mitochondria · mitosis · nucleus · organ · organism · organ system · osmosis · partially permeable membrane · photosynthesis · sexual reproduction · root hair cells · stomata · tissue · vacuole

1.1 Animal cells

Co-ordinated	Modular
10.1	Mod 01 10.1

A study of **organ systems** in humans shows that they are made up of **cells**, **tissues** and **organs**. To show how they are all linked, consider the digestive system (see section 2.1).

- An **organism** is made up of a number of organ systems.

- An organ system is a number of organs linked so that they work together to perform a particular function. In the case of the digestive system, the stomach is just one of the organs involved.

- An organ is made of a number of different tissues, each with a particular role. Different tissues are combined to form the stomach.
  - It has an inner lining of tissue that secretes **enzymes** and mucus.
  - It has muscle tissues that churn the stomach contents by their contraction and relaxation.
  - It has nervous tissue and a supply of blood (a connective tissue).

- Each tissue is made of a group of cells that all have the same structure and are working together to carry out a particular function. In the stomach there are four different groups of cells each working together to make the stomach perform all its functions.

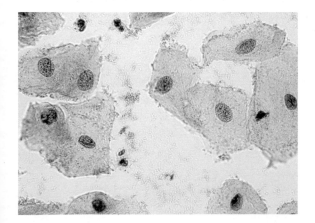

All living things are made up of cells

Figure 1.1

Figure 1.1 provides a summary of the levels of organisation within organisms.

Organism – made of all the
different organ systems

↑

Organ – made up of two or more tissues
e.g. the eye, the heart or a leaf

↑

Tissue – containing similar cells all
performing the same function
e.g. muscle, xylem or phloem

↑

Cell – the building block
for all living things

All living things are made of cells – they are the 'building bricks' of life. There are a huge variety of specially adapted cells carrying out specific functions; for example humans have nerve cells, red blood cells, white blood cells, and skin cells.

Despite the huge range of different cells, most animal cells contain the same parts (see Figure 1.2).

The important parts of a cell are:

- the **nucleus**.
  - it contains the **chromosomes** that carry the **genes** (see sections 1.5 and 5.2). It is the genes which control the characteristics of the cell.
  - it is responsible for controlling all the chemical activities going on in the cell.
  - it controls cell division.

- the **cell membrane** – this is a thin barrier between the cell contents and the outside of the cell. It controls the movement of substances into and out of the cell. This includes the entry of useful chemicals such as water, oxygen and glucose, and the removal of waste chemicals such as carbon dioxide.

Because the cell membrane controls which chemicals pass into and out of the cell, it is described as being a **partially permeable membrane**.

- the **cytoplasm** – this is the substance that fills the space within the cell membrane. All the chemical reactions take place in the cytoplasm.

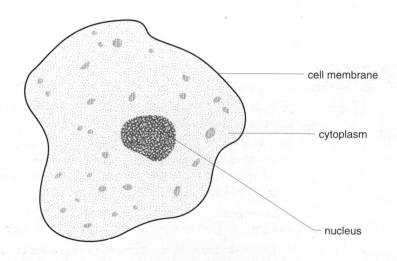

cell membrane

cytoplasm

nucleus

Figure 1.2
A typical animal cell

2

Figure 1.3
A mitochondrion – the site of aerobic respiration in the cell

The chemical reactions are controlled by **enzymes**.

● **mitochondria.** These are found in the cytoplasm and can be described as the 'power station' of the cell, because they are the site of **aerobic respiration** (see section 2.5). Although tiny, their inner membrane is highly folded, providing a very large surface area where the chemical reactions that release the energy needed by the cell take place. Cells that require a lot of energy such as muscle cells and sperm cells have a large number of mitochondria.

Summary

◆ **Organ systems** are made of **organs**.

◆ Each system carries out a particular function or range of functions.

◆ **Organs** are made of **tissues**.

◆ **Tissues** are collections of **cells** working together to carry out a particular function.

◆ Most cells have a **nucleus** which controls the activities of the cell, **cytoplasm** in which chemical reactions take place and a **cell membrane** which controls the passage of substances in and out of the cell.

◆ The chemical reactions in a cell are controlled by **enzymes**.

◆ Energy from respiration is released by the mitochondria in the cytoplasm.

Topic Questions

1 Put the following in order of size, smallest first:
cell   nucleus   organ   organ system   tissue.

2 Complete the gaps in the following table.

Name of part	Function
nucleus	
cytoplasm	
	this controls the passage of substances moving into and out of the cell

3 a) What controls the chemical reactions that take place in a cell?
b) What are mitochondria?

1.2 Plant cells

Co-ordinated	Modular
10.1	Mod 2
	11.1

All plant cells, like animal cells have a nucleus, a cell membrane, cytoplasm and mitochondria. In addition, the following features are found only in plant cells.

● The **cell wall.** This is important in a plant cell because it is a rigid layer, made mostly of **cellulose**, which helps to strengthen the cell. Cellulose allows water and other substances to move freely in and out of the cell. The presence of water in the plant cell also helps to give the cell shape and to support the plant.

● **Chloroplasts** – the site of **photosynthesis**. Chloroplasts are small and disc shaped, containing molecules of the green pigment **chlorophyll** which absorbs light energy.

● Large **vacuoles.** These contain cell sap, which is mostly water and dissolved substances like sugars. Vacuoles and their cell sap also help the plant cell keep its shape and give support to the young plant.

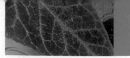

Cell activity

Figure 1.4
A typical plant cell

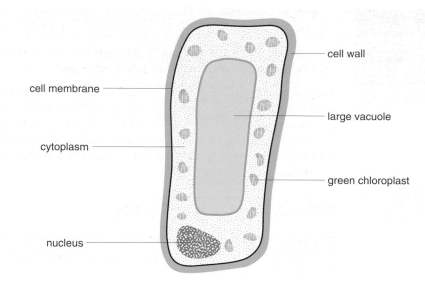

cell wall

cell membrane

large vacuole

cytoplasm

green chloroplast

nucleus

Figure 1.5 lists those parts found in animal and plant cells.

Figure 1.5

Part	Plant cell	Animal Cell
a nucleus	✓	✓
a cell membrane	✓	✓
cytoplasm	✓	✓
mitochondria	✓	✓
a cell wall	✓	✗
a vacuole	✓	✗
chloroplasts	✓	✗

Summary

◆ Plant cells like animal cells have a nucleus, a cell membrane, cytoplasm and mitochondria. In addition most plant cells have **chloroplasts** which absorb light energy to make food, a large **vacuole** which is filled with cell sap, and a **cell wall** which strengthens the cell.

Topic Questions

1 a) What is the name of the green substance found in chloroplasts?
  b) Complete the gaps in the following table.

Name of part	Function
chloroplast	
cell wall	
	filled with cell sap to give support

2 a) List the four parts that both animal and plant cells have.
  b) What three extra parts do plant cells have?

1.3	
Co-ordinated	Modular
10.2	Mod 01
	10.6

Diffusion

Substances enter and leave cells through the cell membrane and cell wall by a number of processes, one of which is **diffusion**.

Diffusion is the movement of particles from a region where there is a high concentration of the particles to where there is a lower concentration of the particles. For example, if a girl sprayed herself with perfume, the perfume particles would move very quickly away from the girl and soon become evenly spread throughout the room. Diffusion is the process that allows the smell of the perfume to spread around the room.

Diffusion in liquids and gases

In a gas, particles move around freely and, like the girl's perfume, will spread out completely within their container. Liquid particles move about less freely within the volume of the liquid.

If a dye is put carefully into a beaker of water, the colour can be seen spreading slowly through the water even if the water is not stirred.

Figure 1.6
Demonstrating diffusion in a liquid

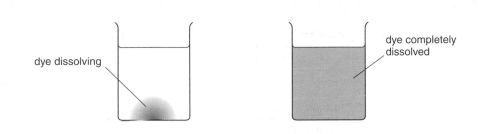

dye dissolving

dye completely dissolved

Similarly, bromine, a dark brown gas, will quickly spread throughout a jar of air until the whole of the inside is a uniform colour. This shows that the air and bromine are completely mixed.

Figure 1.7
Diffusion of bromine vapour

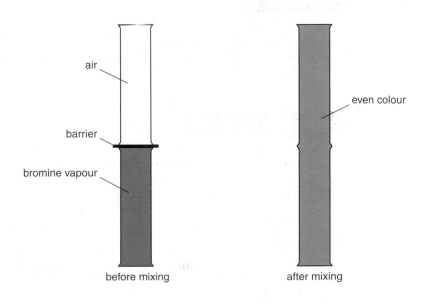

air

barrier

bromine vapour

before mixing

even colour

after mixing

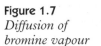
Did you know?

Gas particles move at very high speeds in oxygen gas, a molecule of oxygen has speed of 1000 mph but does not get very far because it has 7000 million collisions a second with other oxygen molecules.

Diffusion in gases is much faster than diffusion in liquids because the particles in a gas move much faster than those in a liquid.

Diffusion in living things

Oxygen and carbon dioxide can move quickly into and out of cells by diffusion. This is because the gases are made up of small molecules. Bigger molecules, such as those of glucose, can also pass through the cell membrane but they take a little longer to do so.

The rate at which substances pass through depends on their concentration (i.e. the number of molecules of each substance) on either side of the membrane – if there is a big difference in concentration (a large concentration gradient), the movement is rapid.

Diffusion does not use any energy. It is the way in which oxygen leaves the alveoli (see section 2.4) and enters the red blood cells and carbon dioxide leaves the blood and goes into the alveoli during gas exchange in the lungs. It is also the way small molecules leave the small intestine and enter the bloodstream during digestion.

Figure 1.8
Diffusion of gases between an alveolus and a blood capillary in the lung

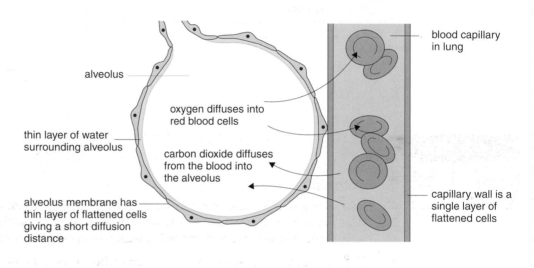

alveolus

blood capillary in lung

oxygen diffuses into red blood cells

thin layer of water surrounding alveolus

carbon dioxide diffuses from the blood into the alveolus

alveolus membrane has thin layer of flattened cells giving a short diffusion distance

capillary wall is a single layer of flattened cells

In plants, water diffuses into the **root hair cells**. During **photosynthesis** carbon dioxide diffuses into the leaf through the **stomata** and oxygen diffuses out through the stomata.

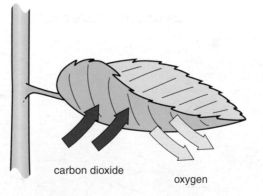

carbon dioxide

oxygen

Figure 1.9
The diffusion of gases into and out of the lower surface of a leaf

Summary

♦ **Diffusion** is the movement of particles from a region where they are at a higher concentration to a region where they are at a lower concentration.

♦ During diffusion the particles move down a concentration gradient.

♦ The greater the difference in concentration the faster the rate of diffusion.

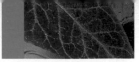

1 What is diffusion?

2 a) Complete the following table.

% concentration of oxygen		In which direction will oxygen move?	Why?
Region A	Region B		
20	5		
5	30		
10	0		

b) In which pair of concentrations will diffusion take place most rapidly? Give a reason.

1.4	
Co-ordinated	Modular
10.2	Mod 02
	11.3

Osmosis and active transport

Osmosis

Osmosis is a special kind of diffusion involving the movement of water molecules. Water molecules are small and can pass easily through a partially permeable membrane that will prevent the movement of larger molecules.

This special kind of diffusion only occurs when a partially permeable membrane separates two solutions. So osmosis is the movement of water through a partially permeable membrane from a region of high water concentration to a region of lower water concentration. The water is said to move down a concentration gradient.

partially permeable membrane

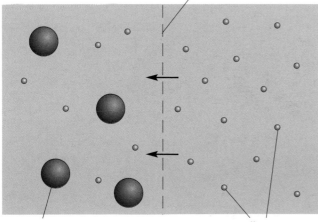

large solute molecule

small water molecules

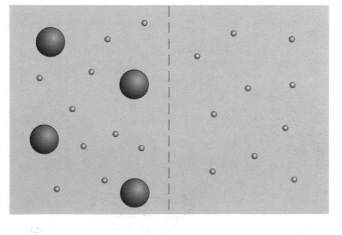

Figure 1.10
A concentration gradient across a partially permeable membrane

Figure 1.11
The water molecules have moved through the partially permeable membrane until there is the same concentration on each side

Cell activity

If a potato chip is put into a beaker containing water (Figure 1.12), then after a while the chip weighs more than it did at first and the chip has become firmer. Water has moved into the chip by osmosis.

But when a similar chip is put into a beaker of sugar solution, after a while the chip has lost weight and is very soft because water has moved out of the chip into the sugar solution by osmosis.

Dialysis tubing can be used in a similar way (see Figure 1.13). Dialysis tubing is a partially permeable membrane.

Osmosis is a very important process. It enables plants to take in water through their roots because there will usually be a lower concentration of water inside the cells of the root hairs than outside in the soil.

As water enters the root hair cell, the contents of the cell become more dilute, so water moves by osmosis to the next cell, which in turn becomes more dilute and so water passes on to the next cell. In this way water reaches the xylem and then travels up the stem by the transpiration pull (see section 4.3). The root hair cell keeps losing water to the next cell and so keeps gaining water by osmosis because the concentration gradient is maintained.

Figure 1.12
Water moves from a region of high water concentration to a region of lower water concentration by osmosis

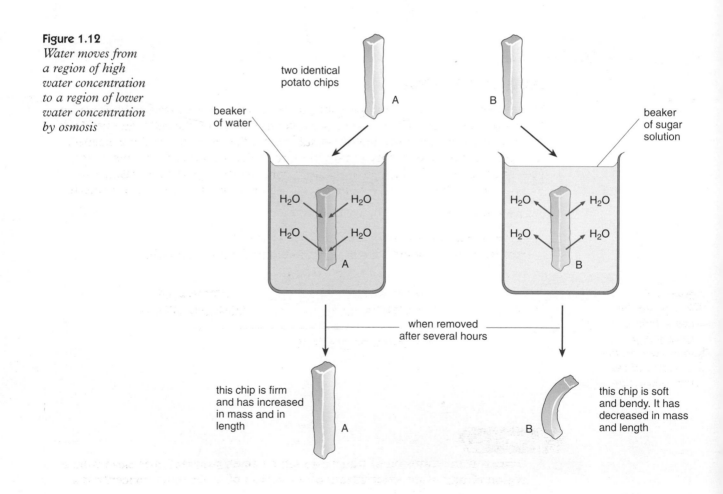

Figure 1.13
The results of an osmosis experiment using dialysis tubing

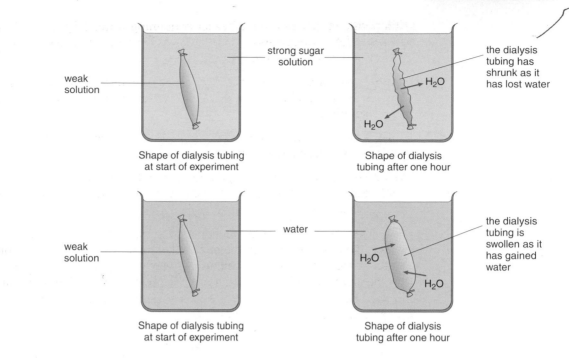

Shape of dialysis tubing at start of experiment

Shape of dialysis tubing after one hour

Shape of dialysis tubing at start of experiment

Shape of dialysis tubing after one hour

Active transport

Sometimes it is necessary to move substances against a concentration gradient. In these situations there will be a greater concentration inside the cell than outside. If the cell needs to take in some of these substances then energy transfer is needed. The cell must transfer energy to take in more of the substance from outside the cell. The taking in of a substance against a concentration gradient is called active transport. It is called active transport because there is an energy transfer involved.

Active transport:

● allows a plant to take in mineral ions from the soil.
● explains why many mitochondria are found in the root hair cells.

Figure 1.14
Water moves by osmosis from a region of high water concentration to a region of low water concentration

HIGH WATER
CONCENTRATION $\Rightarrow$ LOW WATER
CONCENTRATION

Concentration gradient $\longrightarrow$

Summary

◆ **Osmosis** is the diffusion of water through a **partially permeable membrane** from a region of high water concentration to a region of lower water concentration.

◆ The partially permeable membrane allows the passage of water molecules but not solute molecules.

◆ **Active transport** is the absorption of substance against a concentration gradient. This absorption requires the energy from respiration.

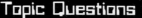

Topic Questions

1 a) What is osmosis?
   b) What is a partially permeable membrane?
   c) What part of a cell is the partially permeable membrane?

2 Put these sugar solutions in order of **increasing** concentration.

Solution	Volume of water (cm^3)	Amount of sugar dissolved (g)
A	100	8
B	100	12
C	50	3
D	50	5
E	200	10
F	200	18

3 Complete the following table.

% concentration of sugar		In which direction will water move?	Why?
Region A	Region B		
10	15		
50	15		
12	12		

4 In what way is active transport different from either osmosis or diffusion?

5 a) What does the cell need to supply if active transport is to occur?
   b) Which living process supplies this need?

6 Why are the cells in root hairs well supplied with mitochondria?

1.5 Cell Division

Co-ordinated	Modular
10.3	Mod 04
	13.1

Mitosis

The nucleus of a cell contains **chromosomes**. Each chromosome contains large numbers of **genes** that control characteristics such as eye colour. In body cells chromosomes are normally found in pairs. Therefore in each cell there will be two genes for each characteristic. Many genes, including the gene for eye colour, have different forms called **alleles**. One allele might be responsible for blue eyes and the other allele responsible for brown eyes.

Body cells divide to produce additional cells during growth or to produce replacement cells. When body cells divide, each cell receives an identical copy of the genetic information of the parent cell.

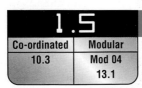

Figure 1.15
A chromosome

The division of body cells is by a process called **mitosis**. During mitosis a copy of each chromosome is made. When the cell divides each new body cell contains identical genetic information.

Some organisms are able to reproduce through mitosis. For example, strawberry plants grow runners which separate into new individuals. The new plants are genetically identical to the parent plant. This is called **asexual reproduction**.

Meiosis

Sexual reproduction involves two parents who each produce sex cells that must be joined together at **fertilisation** (**fusion**) to develop into the new individual.

Meiosis is cell division that leads to the formation of the **gametes**. In meiosis a copy of each chromosome is made. The cell divides twice, so that each cell produced by meiosis has half the number of chromosomes.

In humans meiosis takes place in the testes and ovaries. Each individual has 23 pairs of chromosomes, one of each pair coming from the mother and the other from the father. At the start of meiosis these chromosomes form pairs so that during cell division the gametes formed only have one chromosome from each pair. This process is completely random so all gametes will be different because they will each have an assortment of mother and father chromosomes. Another important point about meiosis is that it keeps the chromosome number the same in all generations with 23 pairs remade when the two sets of 23 single chromosomes in the gametes join at fertilisation.

The important points about meiosis are:

- The gametes formed have half the number of chromosomes of the parent cell. This means that when fertilisation occurs, the number in the new cell will be the same as in the original parent cell.

- The gametes are all genetically different to each other and so the offspring formed by joining two different gametes together will have a unique set of genetic information.

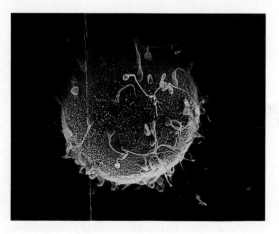

Figure 1.16
A human egg surrounded by sperm. Notice the huge difference in size between the two types of cells

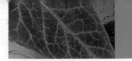

Figure 1.17
*A summary of
sexual reproduction*

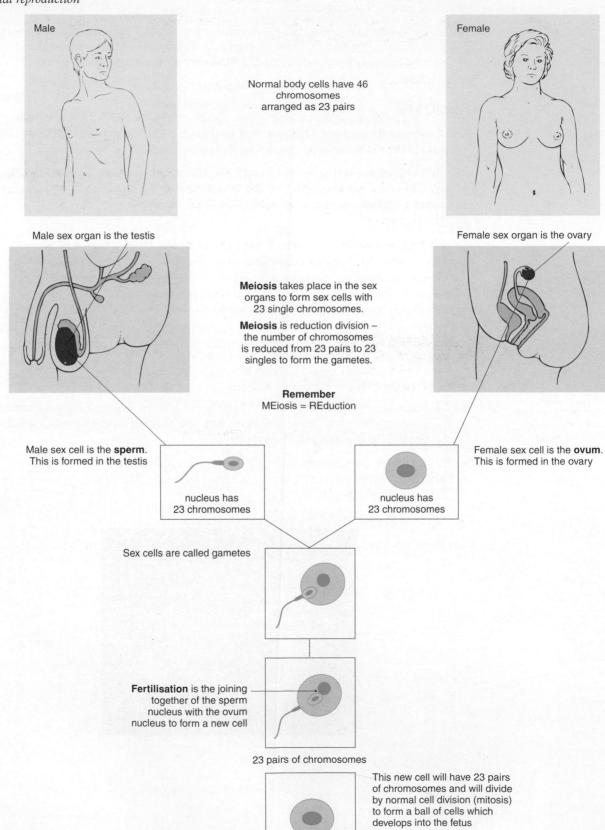

Male

Female

Normal body cells have 46
chromosomes
arranged as 23 pairs

Male sex organ is the testis

Female sex organ is the ovary

Meiosis takes place in the sex
organs to form sex cells with
23 single chromosomes.

Meiosis is reduction division –
the number of chromosomes
is reduced from 23 pairs to 23
singles to form the gametes.

Remember
MEiosis = REduction

Male sex cell is the **sperm**.
This is formed in the testis

Female sex cell is the **ovum**.
This is formed in the ovary

nucleus has
23 chromosomes

nucleus has
23 chromosomes

Sex cells are called gametes

Fertilisation is the joining
together of the sperm
nucleus with the ovum
nucleus to form a new cell

23 pairs of chromosomes

This new cell will have 23 pairs
of chromosomes and will divide
by normal cell division (mitosis)
to form a ball of cells which
develops into the fetus

Summary

◆ A cell nucleus contains **chromosomes**.

◆ Each chromosome carries a large number of **genes** which control the characteristics of the body.

◆ Many genes have different forms called **alleles**.

◆ In body cells the chromosomes are found in pairs.

◆ During **mitosis** a copy of each chromosome in a body cell is made. Cell division takes place and each new body cell contains identical genetic information.

◆ During **meiosis** the cells in the reproductive organs divide to form **gametes**. In this cell division copies of the chromosomes are made, the cell divides twice to form four gametes each with half the number of chromosomes of the parent cell.

◆ During **fertilisation** gametes join to form a single body cell with the complete set of paired chromosomes.

Topic Questions

1   Put these in increasing order of size:
    cell   chromosome   gene   nucleus

2   What are alleles?

3   In a body cell chromosomes are normally found:
    A on their own
    B in pairs
    C in groups of three
    D in groups of four.

4   What happens to a body cell during mitosis?

5   What happens to the cells in the reproductive organs during meiosis?

Examination questions

1 *Amoebae* are single-celled animals. The students took some *Amoebae* from the pond. They looked at one under a microscope.

a) Use the words from the box to name the parts of this *Amoeba*. You may use each word once or not at all. *(3 marks)*

cell membrane	cell wall	chloroplast	cytoplasm
	nucleus	vacuole	

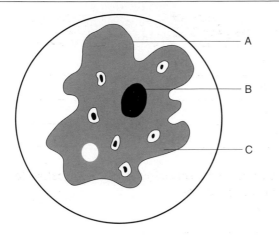

b) Draw lines to link each part of the *Amoeba* cell with its function.

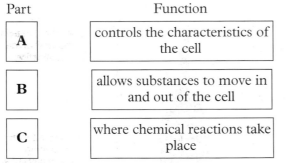

Part | Function

A — controls the characteristics of the cell

B — allows substances to move in and out of the cell

C — where chemical reactions take place

(3 marks)

c) In which of these parts would you find genes?
(1 mark)

2 These diagrams represent types of cells.

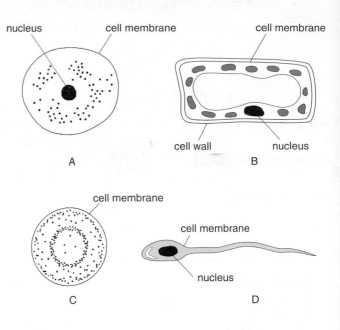

Which diagram A, B, C or D shows a plant cell?
(1 mark)

3 The diagram shows an experiment as it was set up (X) and its appearance one hour later (Y).

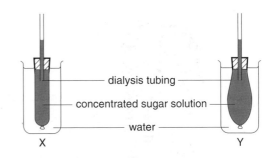

Which is the best explanation for the difference?

A Sugar has moved through the dialysis tubing by osmosis

B Water has been carried into the sugar solution by active transport

C Water has diffused through the dialysis tubing into the sugar solution

D Water has moved from a region of lower water concentration to a region of higher water concentration *(1 mark)*

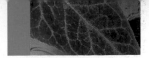

4 The diagrams are experiments set up by a student to study the movement of particles. Test tube 1 shows the start of the experiment. Test tube 2 shows the same tube some time later.

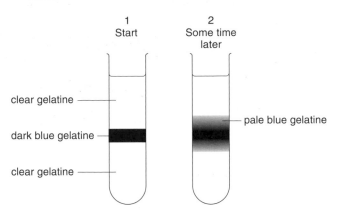

The movement of the blue colour is due to

A active transport
B diffusion
C osmosis
D transpiration

(1 mark)

5 The diagram shows what happens to a plant after 6 hours in a strong salt solution. Why has the plant wilted?

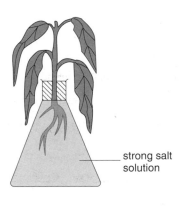

A Salt has entered the plant by diffusion
B Salt has left the plant by osmosis
C Water has entered the plant by diffusion
D Water has left the plant by osmosis

(1 mark)

6 Two chips, P and Q, both 50 mm long, were cut from a potato. P was put into a dish containing water, Q into a dish containing sugar solution.

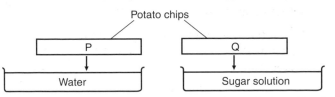

Which row of the table shows the *likely* lengths after one hour?

| | Length of chip in mm | |
	P	Q
A	48	54
B	48	50
C	50	54
D	54	48

7 Some students set up the equipment below to investigate osmosis.

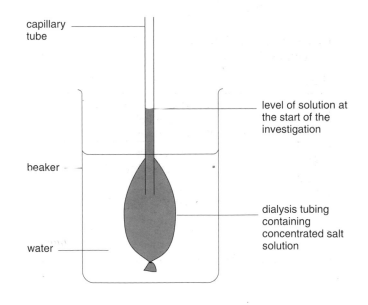

a) What is osmosis? *(3 marks)*
b) i) What will happen to the water level in the capillary tube during the investigation because of osmosis? *(1 mark)*
   ii) Use your knowledge of osmosis to explain why this happens. *(2 marks)*

8 In the cell shown in the diagram as a box, one chromosome pair has alleles **Aa**. The other chromosome pair has alleles **Bb**. The cell undergoes meiosis.

a) Copy and complete the diagram of the four gametes to show the independent assortment, or reassortment, of genetic material during meiosis. *(2 marks)*

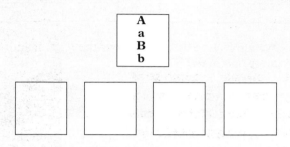

b) If the cell undergoes mitosis instead of meiosis, draw the two daughter cells which result to show the chromosomes in each. *(2 marks)*

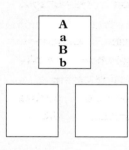

c) State the number of chromosomes in:

i)   a normal human cell   *(1 mark)*
ii) a human gamete   *(1 mark)*
iii) the daughter cell from mitosis of a human cell.   *(1 mark)*

9 The drawing shows part of a root hair cell.

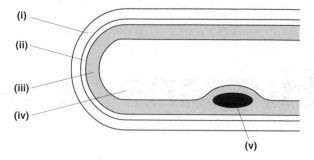

Use words from the list to label the parts of the root hair cell.

**cell membrane   cell wall   cytoplasm
nucleus   vacuole**

10 a) The diagram shows four ways in which molecules may move into and out of a cell. The dots show the concentration of molecules.

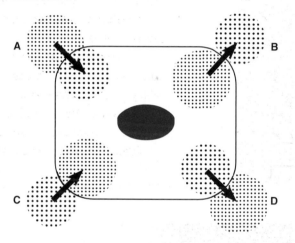

The cell is respiring aerobically.
Which arrow, **A**, **B**, **C** or **D**, represents:

i)   movement of oxygen molecules
ii) movement of carbon dioxide molecules?

(2 marks)

b) Name the process by which these gases move into and out of the cell.

(1 mark)

c) Which arrow, **A**, **B**, **C** or **D**, represents the active uptake of sugar molecules by the cell? Explain the reason for your answer

(2 marks)

Chapter 2
Humans as organisms

Key terms

absorption · aerobic respiration · alveoli · amino acids · amylase · anaerobic respiration · anus · aorta · artery · atria · bile · breathing · bronchi · bronchioles · cilia · capillaries · catalysts · denatured · deoxygenated blood · diaphragm · digestion · emulsifying · enzyme · exhale · faeces · fatty acids · gall bladder · gaseous exchange · glycerol · glycogen · gullet · haemoglobin · heart · inhale · insoluble · lactic acid · large intestine · lipase · lipids · lungs · liver · mitochondria · mucus · nutrition · oesophagus · oxygenated blood · oxygen debt · oxyhaemoglobin · pancreas · pH · plasma · platelets · proteases · pulmonary artery · pulmonary vein · red blood cells · respiration · respire · rib muscles · saliva · salivary glands · small intestine · soluble · stomach · substrate · thorax · tissue fluid · trachea · urea · vein · vena cava · ventilation · ventricles · villi · white blood cells

2.1 Nutrition

Co-ordinated	Modular
10.4	Mod 01
	10.2

Unlike plants, that make their own food by photosynthesis, most animals take **insoluble** food in and then break it down into small, **soluble** molecules that can be absorbed into the blood. This process is called **digestion**. In mammals and many other animals, once the food has been broken down it is **absorbed** into the bloodstream. Digestion is made up of three stages.

Figure 2.1
The three stages of digestion

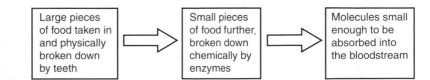

| Large pieces of food taken in and physically broken down by teeth | → | Small pieces of food further broken down chemically by enzymes | → | Molecules small enough to be absorbed into the bloodstream |

What happens during digestion?

The mouth is at the start of the process of digestion. In the mouth, the food is broken down into small pieces and mixed with a special chemical called **saliva**. The teeth break the food down into small enough pieces for swallowing and to allow the **enzyme amylase** to reach a large surface area of the food as quickly as possible. Amylase is made by the **salivary glands** which produce the saliva. The saliva enters the mouth as food is chewed, making the food moist enough to swallow. The amylase in the saliva starts to break the large starch molecules into smaller sugar molecules.

Figure 2.2
The human digestive system

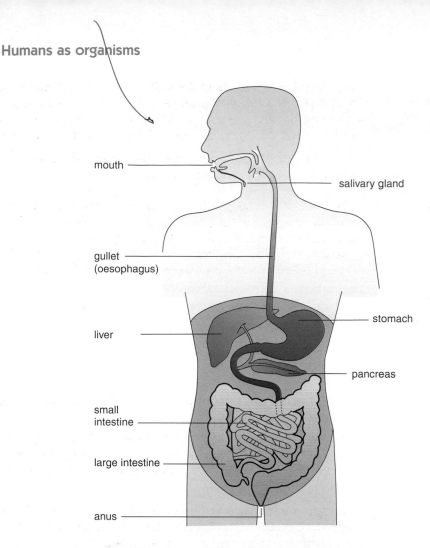

mouth

salivary gland

gullet (oesophagus)

liver

stomach

pancreas

small intestine

large intestine

anus

The mouth is the opening of a long tube. This long tube varies in width as it passes through the body to an opening at the other end called the **anus**. Food does not really enter the body until it passes through the walls of this tube and enters the bloodstream. Some of the food material that we eat, in particular fibre, never enters the body but passes out of the body as **faeces** through the anus.

The **gullet (oesophagus)** is the part of the tube which carries the food from the mouth to the stomach. It is a muscular tube and by repeated contractions, the food is pushed down towards the stomach. This movement through the digestive system is called peristalsis.

Figure 2.3
Peristalsis

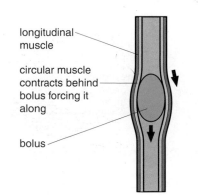

longitudinal muscle

circular muscle contracts behind bolus forcing it along

bolus

The **stomach** is a muscular bag. It has three layers of muscles that work in different directions so that when they contract, it means that the contents of the stomach are squeezed and moved in all directions. As a result, the contents are very well mixed with the gastric juices made in the stomach.

Gastric juices are made in the walls of the stomach and include dilute hydrochloric acid and enzymes called **proteases**. Hydrochloric acid helps to destroy bacteria; it also activates the protease and provides the right pH for the enzyme to work. The proteases are enzymes that break down proteins into **amino acids**.

The contents of the stomach are squirted into the **small intestine**. The small intestine is a long tube, coiled to take up less space.

The contents of the **gall bladder** – the **bile** – and the pancreatic enzymes are emptied into the first part of the small intestine through a small tube from the **pancreas**. Bile does not contain any enzymes but is very important because it breaks down (**emulsifies**) the fats into very small droplets giving them a larger surface area so that the fat digesting enzymes (**lipases**) can work more quickly and efficiently.

Figure 2.4
Bile breaking a fat molecule down into small droplets

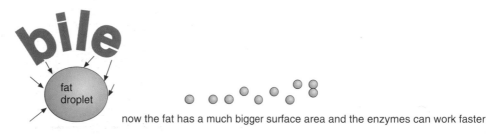

now the fat has a much bigger surface area and the enzymes can work faster

Glands in the wall of the small intestine produce amylase, protease and lipase enzymes that complete digestion. Proteins are digested into amino acids. The remaining starches are digested into glucose, fats are digested into **fatty acids** and **glycerol**.

Another important feature of the small intestine is that it is here that the small digested molecules are absorbed into the blood. The walls of the small intestine are lined with finger-shaped projections called **villi**.

As can be seen in Figure 2.5, the villi provide an efficient absorption surface because they have:

- a large surface area
- a moist surface
- a plentiful blood supply
- a thin membrane.

These features ensure that molecules can be absorbed rapidly into the bloodstream.

Figure 2.5
a) A cross section of the small intestine. b) Villi lining the small intestine

a)

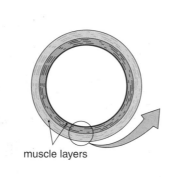

muscle layers

b)

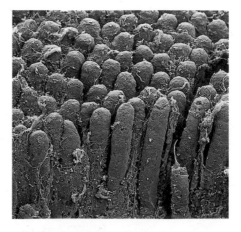

By the time the contents of the small intestine have been moved along to the **large intestine**, the small soluble molecules have been absorbed. The large intestine contains the material that cannot be digested and absorbed by our gut.

This is mostly the fibre in our diet. The fibre is, however, important in providing the bulk of the contents of the intestine, something for the muscles to push against, and helping to move the food through from the mouth to anus.

There is still one very important function to be done before the unusable contents of the large intestine can be removed as faeces. The water, which has made the 'food' soft and easy to push through the digestive system, is now reabsorbed through the lining of the large intestine back into the body. When the faeces are released we are said to defaecate.

Figure 2.6
Summary of digestion

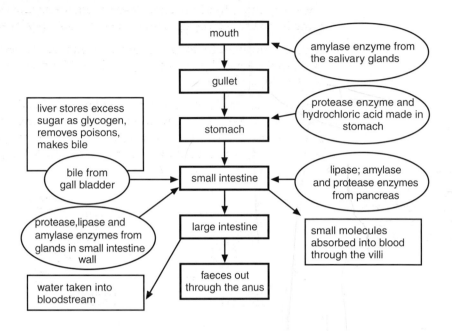

The liver

The **liver** is the largest organ in a mammal and only two of the many important jobs that it does will be considered here. Bile is made in the liver and stored in the gall bladder.

Bile is vital because it contains chemicals that alter the **pH** of the stomach contents as it enters the small intestine. Enzymes in the small intestine operate best at a higher pH. Remember that the stomach contains hydrochloric acid, so the bile salts must be alkaline.

Bile also changes the fats into smaller droplets to provide a larger surface area for the enzymes to work on (Figure 2.4).

Did you know?

One of the other important functions of the liver is the storage of glucose as **glycogen**. Glucose is very important in the process of respiration, so it is essential for the body to be able to obtain it quickly. However glucose alters the osmotic effect of a solution, so if the glucose was stored in the blood, the concentration in the blood would fluctuate constantly. To prevent this, but to be able to call on glucose reserves quickly, the body stores the 'spare' glucose as glycogen in the liver (and muscles).

One way that too much alcohol can damage an individual's health is because alcohol is a poison to the body. The liver cells are damaged as they break down the poisons from our blood.

Digestion by enzymes

The teeth have already started to break down the food by the time it reaches the stomach but the pieces of food are still too big for the body to absorb. The next stage of digestion is that done by enzymes. Enzymes are **catalysts** which speed up the breakdown of large molecules into smaller molecules. There are three main enzymes involved in the digestion of starch, protein and fats.

Enzymes are chemicals with very specific tasks. The enzyme amylase can only break down the large starch molecules to smaller sugar molecules. Amylase cannot break down any other food molecule and no other enzyme can break down starch. The same is true for all the enzymes: they can only perform one job.

Protease enzymes break down proteins into amino acids, and lipases break fats down into fatty acids and glycerol (Figure 2.7). The molecule which reacts with an enzyme is called its **substrate**.

The specific nature of enzymes is explained in section 12.4. It is related to the particular shapes of the enzyme molecules. Overheating an enzyme disturbs the shape so that the substrate no longer fits with the enzyme. Such enzymes are described as having been **denatured** and will not work properly.

Figure 2.7
Enzymes break down starch, protein and fat during digestion

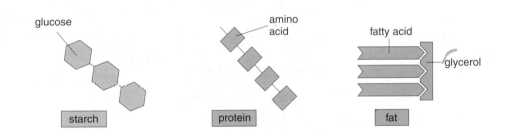

Figure 2.8

Enzyme	Substrate	Product
amylase	starch	sugars
proteases	protein	amino acids
lipase	lipids	fatty acids + glycerol

The sugars, amino acids, fatty acids and glycerol are all small enough to be absorbed through the gut wall into the bloodstream and then to travel to the cells that need them. So the enzymes have completed the task of digestion.

Did you know?

The body has a special way of preventing proteases from digesting its own body proteins. The proteases are not activated (switched on) until there is food present to digest.

Pepsin, the protease made in the stomach, has an optimum pH of 2, but most enzymes work best at pH 7.

Did you know?

The importance of pH to the digestion of starch
Starch makes up much of the diets of people throughout the world. Starch is an example of a group of foods called carbohydrates. The digestion of carbohydrates, such as starch, starts in the mouth through the action of amylase. This causes some of the large starch molecules to be broken down into smaller sugar molecules. However, the action of amylase ceases in the acidic conditions in the stomach and does not continue until the undigested starch reaches the alkali conditions in the small intestine. Here, the amylase produced by the pancreas causes the breakdown of the remaining starch into sugar. It is in the small intestine that another enzyme, maltase, causes the breakdown of the sugar molecules into even smaller molecules of glucose. It is this glucose that is important in respiration.

Summary

- **Digestion** is the process of breaking down large **insoluble** food molecules (such as starch, protein and fats) into smaller **soluble** molecules.

- **Absorption** is the movement of the small soluble molecules from the digestive system into the bloodstream.

- The absorption of the soluble substances takes place through the wall of the **small intestine**.

- The walls of the small intestine are lined with **villi** which are adapted to be efficient absorbers.

- Water is absorbed into the bloodstream in the **large intestine**.

- **Faeces** are indigestible food which leaves the body through the **anus**.

- **Enzymes** are biological **catalysts** that speed up the breakdown of the large molecules.

- **Amylase** (an enzyme produced in the **salivary glands**, **pancreas** and **small intestine**) speeds up the breakdown of starch into sugars.

- **Protease** enzymes (produced in the **stomach**, pancreas and small intestine) speed up the breakdown of protein into **amino acids**.

- **Lipase** enzymes (produced in the pancreas and small intestine) speed up the breakdown of **lipids** (fats and oils) into **fatty acids** and **glycerol**.

- The stomach produces hydrochloric acid which kills bacteria and produces the acidic conditions in which the enzymes in the stomach work most effectively.

- The **liver** produces **bile** which neutralises the acid added in the stomach. It produces the alkaline conditions in which the enzymes in the small intestine work most effectively.

- Bile breaks up (**emulsifies**) large fat drops into small droplets thereby increasing the surface area for enzymes to act on.

Topic questions

1 What is digestion?

2 Where are the breakdown products of digestion absorbed into the bloodstream?

3 How are villi adapted to ensure that the absorption of digested food is as rapid as possible?

4 Through which part of the digestive system is water reabsorbed into the bloodstream?

5 Copy and complete the following table.

6 Give two reasons why hydrochloric acid is produced in the stomach.

7 a) Which organ produces bile?
  b) Explain why bile is important in the process of digestion.

Enzyme	Where produced	Substrate	Products
amylase			
proteases			
lipase			

2.2	
Co-ordinated	Modular
10.5	Mod 01
	10.4

Composition of the blood

Plasma

Plasma is the main transporting fluid in blood. It is a straw-coloured liquid but because it carries so many red cells, blood is red.

Plasma is an important transport medium. It carries

- **red blood cells, white blood cells** and **platelets**

- antitoxins and antibodies (see section 3.5)

- the soluble end products of digestion (amino acids, glucose, fatty acids and glycerol) from the small intestine to other organs

- carbon dioxide from the organs to the lungs

- **urea** from the liver to the kidneys (see section 3.4)

- mineral salts

- hormones (see section 3.2)

- heat, which is distributed around the body in the blood. This helps to regulate body temperature

Red blood cells

Red blood cells are very numerous. Their function is to transport oxygen to the organs.

Figure 2.9
Red blood cells as seen in an electron micrograph

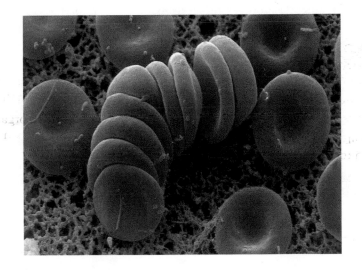

Mature red blood cells do not have a nucleus. This causes them to have a (biconcave) double-dish shape.

This shape is important as it gives

- more packing space for haemoglobin for transporting oxygen

- a large surface area for oxygen to diffuse into the red blood cell

- a short distance for the oxygen to diffuse through to combine with the haemoglobin to form oxyhaemoglobin.

The diameter of a red blood cell is similar to the diameter of a small capillary. But as the cell is surrounded by a flexible membrane, its shape can distort and it can squeeze through the smallest capillaries.

Red blood cells and oxygen transport

The only place in the body that the red blood cells can collect oxygen is the alveoli of the lungs. The oxygen dissolves in the moist lining of the alveoli and diffuses along the concentration gradient into the blood capillaries (see section 1.3). The oxygen then passes by diffusion into the red blood cells and combines with haemoglobin to form oxyhaemoglobin. The blood is now oxygenated and returns in the pulmonary vein to the heart. The heart pumps the blood around the body via the arteries. All tissues need oxygen, especially respiring muscles, and in areas such as these where there is a shortage of oxygen, the oxyhaemoglobin quickly breaks down giving free oxygen that diffuses into the cells where it will be used for respiration.

White blood cells

Some white blood cells fight against infection by ingesting and destroying bacteria (see section 3.5). White blood cells contain a nucleus.

Platelets

Platelets are cell fragments. They play a very important role in the formation of blood clots (see section 3.5). Platelets do not have a nucleus.

Summary

- The fluid part of the blood is called **plasma**.

- Plasma contains **white blood cells, platelets** and **red blood cells**.

- Plasma transports:
  - carbon dioxide from the organs to the lungs
  - soluble end-products of digestion from the small intestine to other organs
  - **urea** from the liver to the kidneys.

- **White blood cells** have a nucleus and are part of the body's defence system.

- **Red blood cells** transport oxygen from the lungs to the organs.

- **Platelets** are cell fragments that help blood to clot.

- **Red blood cells** have no nucleus and contain haemoglobin.

- In the lungs oxygen joins with haemoglobin to form oxyhaemoglobin.

- In organs, where the concentration of oxygen is low, oxyhaemoglobin releases oxygen and becomes haemoglobin.

Topic questions

1 What is plasma?

2 Which blood cells contain a nucleus?

3 a) Why do red blood cells have a large surface area?
   b) Explain why it is important that red blood cells have a flexible membrane.

4 a) Which cells take oxygen from the lungs to the organs of the body?
   b) Which part of the blood takes carbon dioxide from the organs to the lungs?

5 What is the function of white blood cells?

6 a) What are platelets?
   b) What is their function?

7 a) What is the name of the substance that gives red blood cells their colour?
   b) Where and when is oxyhaemoglobin formed?
   c) What happens to oxyhaemoglobin in organs which respire actively (e.g. the liver)?

2.3	
Co-ordinated	Modular
10.5	Mod 01
	10.4

The circulatory system

Figure 2.10 shows the important parts of the circulatory system.

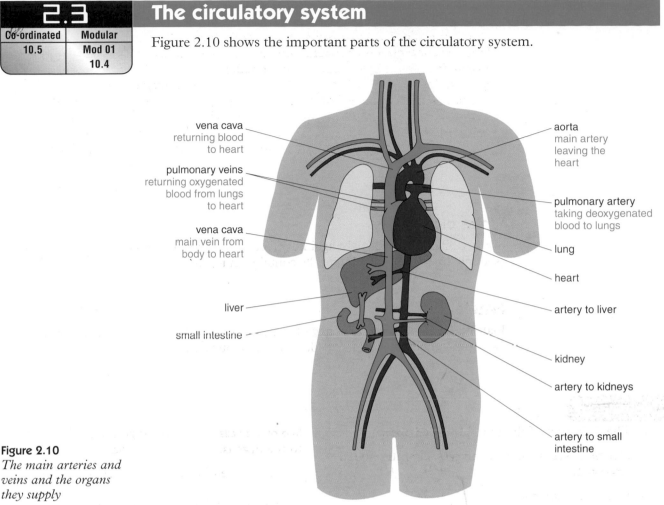

vena cava
returning blood
to heart

pulmonary veins
returning oxygenated
blood from lungs
to heart

vena cava
main vein from
body to heart

liver

small intestine

aorta
main artery
leaving the
heart

pulmonary artery
taking deoxygenated
blood to lungs

lung

heart

artery to liver

kidney

artery to kidneys

artery to small
intestine

Figure 2.10
*The main arteries and
veins and the organs
they supply*

The heart

The **heart** is a four-chambered, muscular, double pump. One part pumps blood to and from the lungs, and the other part pumps blood around the body and back to the heart.

The two top chambers are the **atria** (left atrium and right atrium) – these receive the blood at low pressure from veins. The lower chambers are the pumping chambers called **ventricles**. There is a division between the two sides of the heart called the septum.

The heart is a double pump. The right ventricle pumps **deoxygenated blood** to the lungs (through the **pulmonary artery**) where it absorbs oxygen and returns to the heart via the **pulmonary vein**. The left ventricle pumps **oxygenated blood** through the **aorta** and around the body to the organs and tissues. Blood returns to the heart via the **vena cava**.

Figure 2.12 shows a human heart from the outside. The first artery to come from the aorta is the coronary artery. This supplies oxygenated blood to the heart muscles. The heart muscles are contracting all the time and so need a constant supply of oxygen and nutrients. If the coronary artery gets blocked, either by a fatty deposit or a blood clot, the shortage of oxygen to the muscles causes severe cramp-like pain – a 'heart attack'.

Oxygen is essential for living cells for respiration, and because organs carry out vital functions, each organ must receive a supply of oxygenated blood.

Did you know?

Babies born with a 'hole in the heart' have a gap in the septum. This sometimes grows together naturally but if it is large it may need surgical repair.

25

Figure 2.11
Blood flow through the heart

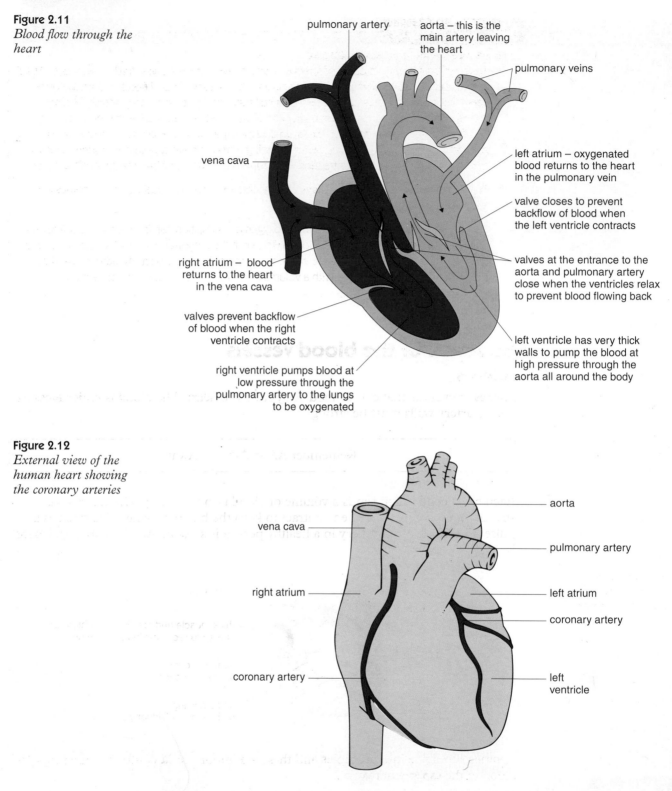

pulmonary artery

aorta – this is the main artery leaving the heart

pulmonary veins

vena cava

left atrium – oxygenated blood returns to the heart in the pulmonary vein

valve closes to prevent backflow of blood when the left ventricle contracts

valves at the entrance to the aorta and pulmonary artery close when the ventricles relax to prevent blood flowing back

right atrium – blood returns to the heart in the vena cava

valves prevent backflow of blood when the right ventricle contracts

right ventricle pumps blood at low pressure through the pulmonary artery to the lungs to be oxygenated

left ventricle has very thick walls to pump the blood at high pressure through the aorta all around the body

Figure 2.12
External view of the human heart showing the coronary arteries

vena cava

right atrium

coronary artery

aorta

pulmonary artery

left atrium

coronary artery

left ventricle

Did you know?

- Ventricles pump out 60 cm^3 of blood in each beat.
- The human heart has a mass of 0.4% of body mass.
- The heart and blood vessels contain about 5 dm^3 blood in an adult male.

You can work out your blood volume using the rough rule of 1 dm^3 for each 10 kg body mass.

Heart beat, heart sounds and pulse

- The heart beat starts in a patch of special muscle tissue in the right atrium which acts as a 'natural pacemaker'. The beat spreads over the atria and then via specialised fibres to the base of the ventricles. Both ventricles contract at the same time, starting at the bottom and squeezing the blood up into the arteries. The sounds of the heart beat are caused by the valves closing. The first sound comes from the bicuspid and tricuspid valves, and the second sound is formed as the semi-lunar valves (at the opening of the aorta and pulmonary artery) close at the end of the ventricular contraction. The pulse is a measure of the rate the heart is beating.

- The fetal heart starts beating 21 days after fertilisation. This first beating cell becomes the 'pacemaker'.

- During vigorous aerobic exercise a man's oxygen consumption can increase 20-fold. The heart output is related to oxygen consumption. To match the increased demand for oxygen by the muscles, there must be an increase in rate and force of the heart beat. An athlete's heart can pump out 40 dm^3 per minute with a volume of 200 cm^3 per ventricular contraction.

Structure of the blood vessels

Arteries

Arteries are vessels that carry blood away from the heart. The blood is under pressure and so artery walls must be strong.

Remember AA = Arteries Away

Each heart contraction sends a volume of blood into the artery. The artery walls stretch to receive this and then contract to keep the blood moving. This is felt as a pulse. The lining of the artery in a healthy person is smooth and the lumen (the bore) is small.

Figure 2.13 ▶
Transverse section through an artery

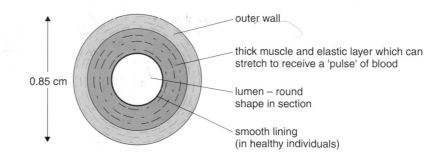

Arteries subdivide into arterioles and these are important in controlling the supply of blood to the **capillary** networks.

Figure 2.14 ▶
Transverse section through a capillary

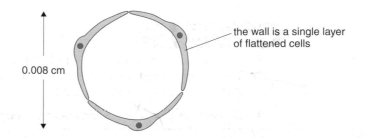

The capillary networks

Although **capillaries** are the smallest of the blood vessels, they are very important as they run close to all cells. Capillaries have walls made of a single layer of flat cells and they have a small lumen.

Plasma can pass out of capillaries to form the **tissue fluid**. Tissue fluid is a watery solution containing all the dissolved substances from the plasma that bathes the cells.

The cells take up substances (such as oxygen and glucose) from the tissue fluid and pass waste products (such as carbon dioxide and urea) into the tissue fluid along concentration gradients.

Figure 2.15 ▶
Exchange between the blood and the tissue cells at the capillary network

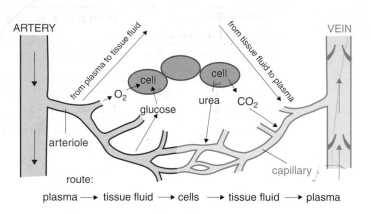

route:

plasma → tissue fluid → cells → tissue fluid → plasma

Veins

Veins receive blood from the capillary network and so carry blood at low pressure. The movement of the blood is helped by contraction and tone of the skeletal muscles and by pocket valves that prevent back-flow.

Veins join to the vena cava and return blood to the heart.

Some veins run close to the surface of the skin such as those which can be seen on the forearm and back of the hand.

Figure 2.16
Longitudinal section (a) and transverse section (b) through a vein

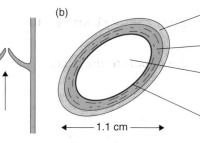

(a) pocket valve allows blood to flow in one direction only

(b)
outer wall – thinner than artery wall

thin layer of muscle and elastic

lumen – larger than in artery and irregular in shape

smooth lining

◄— 1.1 cm —►

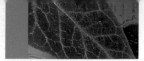

Summary of the transport functions of the blood

Transport of		
Soluble products of digestion	from villi	to all cells
Dissolved mineral salts and ions	from villi	to all cells
Oxygen in red blood cells (as oxyhaemoglobin)	from alveoli in lungs	to all respiring cells
Carbon dioxide (dissolved in plasma)	from respiring cells	to the alveoli of the lungs
Urea in solution in the plasma	from cells	to kidneys
Hormones in solution in the plasma	from endocrine glands (see section 3.2)	to target organs
Heat energy	from active muscles and the liver	to the skin capillaries, nasal capillaries and alveolar capillaries

Summary

◆ The circulatory system consists of the **heart, arteries, veins** and **capillaries.**

◆ The heart is a four-chambered muscular pump that pumps blood under pressure around the body.

◆ There are two separate circulatory systems:
  – one linking the heart and the lungs,
  – the other linking the heart with all the other organs.

◆ Blood flows from the heart through arteries to the organs.

◆ Blood returns from the organs to the heart through veins.

◆ Arteries have thick walls containing muscle and elastic fibres.

◆ Veins have thin walls and often have valves to prevent the back-flow of blood.

◆ Capillaries are very narrow blood vessels with very thin walls.

◆ Substances needed by the cells pass out of the blood through the walls of the capillaries.

◆ Substances produced by the cells pass into the blood through the walls of the capillaries.

Topic questions

1 In which direction does blood flow in:
  a) an artery
  b) a vein?

2 Describe the walls of:
  a) arteries
  b) veins
  c) capillaries.

3 The thickest muscular wall in the heart is that of the left ventricle. Why is this wall so thick?

4 The heart can be considered to be a double pump. Explain why?

5 a) What is tissue fluid?
  b) Why is tissue fluid important?

6 The veins and the heart contain valves. Why?

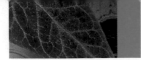

2.4

Co-ordinated	Modular
10.6	Mod 01
	10.3

Breathing

Breathing, **gaseous exchange** and **respiration** are not the same thing.

- Breathing is the process that moves air in and out of the lungs, also known as **ventilation**.
- Gaseous exchange is the exchange of oxygen and carbon dioxide at the lung surface.
- Respiration is the chemical process that takes place in each living cell which transfers energy from food molecules.

Breathing is important as it enables us to obtain oxygen and get rid of carbon dioxide. There is a difference between the inhaled (breathed in) air and the exhaled (breathed out) air. Figure 2.17 shows this difference.

Figure 2.17

	% composition inhaled air	% composition exhaled air
oxygen	20.9	16.3
carbon dioxide	0.04	4.1
nitrogen	79.0	79.5

Oxygen is removed from the air which is breathed into our lungs. The oxygen passes into the bloodstream for use in respiration. The oxygen is replaced by carbon dioxide produced by respiration. The exchange of oxygen and carbon dioxide in the lungs is called gaseous exchange.

Figure 2.18 ▼

The structure of the breathing system in humans

The breathing system

Our gas exchange organs are called the **lungs**. They are part of the breathing system, which is shown in Figure 2.18.

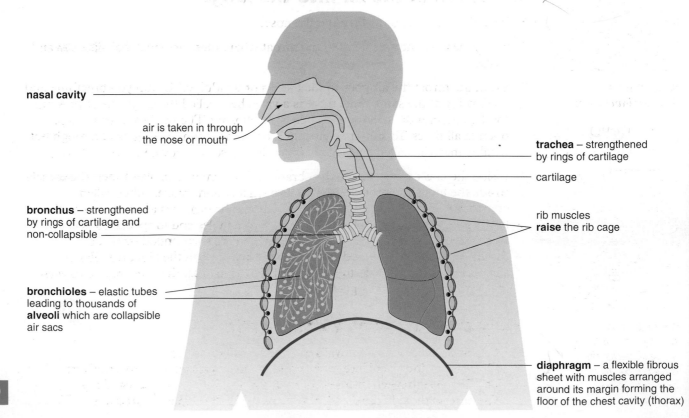

nasal cavity

air is taken in through the nose or mouth

trachea – strengthened by rings of cartilage

cartilage

bronchus – strengthened by rings of cartilage and non-collapsible

rib muscles **raise** the rib cage

bronchioles – elastic tubes leading to thousands of **alveoli** which are collapsible air sacs

diaphragm – a flexible fibrous sheet with muscles arranged around its margin forming the floor of the chest cavity (thorax)

We have two lungs inside the chest or **thorax**. The thorax is bounded and protected by the **ribs** and between the ribs are the **rib muscles**. Below the lungs is a dome-shaped sheet of smooth muscle which separates the thorax from the abdomen. This is called the **diaphragm**. The lungs are made airtight by two very thin sheets of tissue. A lubricating fluid helps the lungs to move smoothly inside the thorax. The lungs appear to be spongy but under the microscope they can be seen to consist of cavities (**alveoli**) and tubes through which air flows into and out of them.

Figure 2.19 ▶
Blood supply to the alveoli

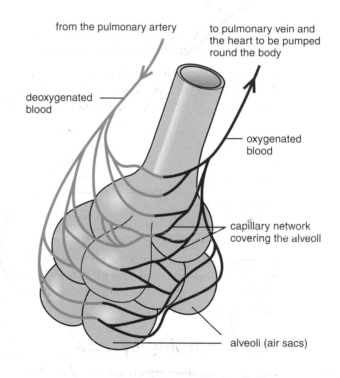

from the pulmonary artery

to pulmonary vein and the heart to be pumped round the body

deoxygenated blood

oxygenated blood

capillary network covering the alveoli

alveoli (air sacs)

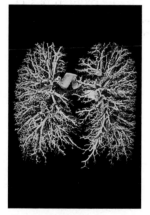

Figure 2.20 ▲
The respiratory tree

Figure 2.21 ▲
Micrograph of lung tissue showing the alveoli and their associated blood capillaries

The route of the air into the lungs

1   The air enters the body through the nose.

2   From the nose the air passes to the throat where there are small hair-like **cilia** and **mucus**.

3   From the throat the air goes to the **trachea** or windpipe. To stop you breathing and swallowing at the same time, there is a valve here called the epiglottis. It prevents food going into the windpipe and causing choking. The trachea must be kept open at all times. To stop it collapsing it has almost complete rings of a tough but flexible material called cartilage in its walls. These are shown in Figure 2.18.

4   Inside the chest cavity the trachea branches to form two smaller tubes, the **bronchi**. Inside the lungs the bronchi branch into smaller and smaller tubes called **bronchioles**. Figure 2.20 shows the bronchi and bronchioles after the rest of the lung tissue has been dissolved away. The bronchioles end in special bags called alveoli. The alveoli have very thin walls, which are surrounded by blood capillaries. Figures 2.19 and 2.20 show the alveoli with the blood capillaries surrounding them. It is at the surface of the alveoli that the exchange of oxygen and carbon dioxide takes place.

How are the gases exchanged?

- The alveoli have very thin walls as shown in Figure 2.21.
- The walls of the alveoli are moist and the oxygen dissolves in this moisture.
- The dissolved oxygen diffuses through the wall into the blood capillary.
- The carbon dioxide diffuses from the blood in the opposite direction.

31

- To make sure that there is always fresh air with lots of oxygen inside the lungs, air must be changed every few seconds. This is done during breathing.

The alveoli are well adapted for gaseous exchange. They have:

- a large surface area
- a moist surface
- a very thin membrane
- a rich capillary network.

Breathing

This has two parts:

1. Inhaling – when the rib cage is raised and the diaphragm contracts and flattens.

   This increases the volume of the chest cavity and lowers the pressure so air rushes into the lungs.

2. Exhaling – when the rib cage is lowered and the diaphragm raised.

   The volume of the thorax is reduced, the pressure inside is raised and the air is forced out. Figure 2.22 shows how this happens.

Breathing takes place automatically. Breathing is controlled in the brain where there are special sense cells that can detect any changes in the amount of carbon dioxide in the blood. If the level of carbon dioxide begins to rise, the brain makes us breathe more quickly and more deeply to increase the amount of oxygen and reduce the amount of carbon dioxide in the body. Carbon dioxide can be very poisonous if it is allowed to build up in the body.

Summary

- During the process of **gaseous exchange** oxygen is removed from the air and diffuses into the bloodstream and carbon dioxide diffuses out of the bloodstream into the air.

  - The **alveoli** are well adapted for **gaseous exchange** with their very large, moist surfaces and their rich supply of blood capillaries.

- The action of the ribcage and diaphragm causes air to pass into and out of the lungs. The movement of air is called **breathing** or **ventilation**.

- Inhaling causes the volume of the chest cavity to increase, so lowering the pressure causing air to enter the lungs.

- Exhaling causes the volume of the chest cavity to decrease, so increasing the pressure causing air to be forced out of the lungs.

Topic questions

1. What happens to the ribcage and the diaphragm:
   a) when you breathe in?
   b) when you breathe out?

2. During breathing what gas:
   a) diffuses from the air into the bloodstream?
   b) diffuses from the bloodstream into the air?

3. Describe as fully as possible what causes the air to pass into your lungs when you breathe in.

4. In what ways is the structure of the alveoli adapted to ensure the rapid diffusion of gases?

Figure 2.22
The breathing action

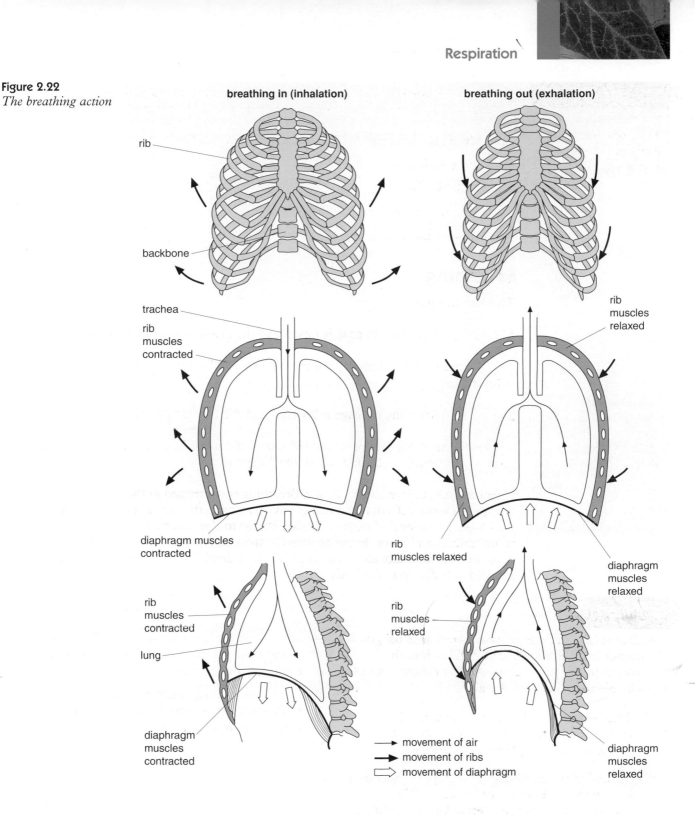

breathing in (inhalation)   breathing out (exhalation)

rib

backbone

trachea
rib
muscles
contracted

rib
muscles
relaxed

diaphragm muscles
contracted

rib
muscles relaxed

diaphragm
muscles
relaxed

rib
muscles
contracted

lung

diaphragm
muscles
contracted

rib
muscles
relaxed

diaphragm
muscles
relaxed

→ movement of air
➡ movement of ribs
⇨ movement of diaphragm

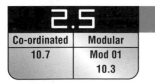

2.5		Respiration
Co-ordinated	Modular	
10.7	Mod 01	
	10.3	

Respiration

Respiration is a little like burning fossil fuels – oxygen is used to enable the transfer of energy from the fossil fuels in the form of heat. At the same time, water and carbon dioxide are released.

A similar chemical reaction is happening in every living cell. To release energy, cells use the oxygen that is breathed in and the glucose from the food that is eaten. Water and carbon dioxide are given off as waste products.

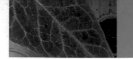

Which cells respire the most? To answer that question it is necessary to think about the cells which need the most energy. Muscle cells and sperm cells are examples. Cells in the roots of plants where minerals are taken in by active transport also need a lot of energy so they respire more than other plant cells (see section 1.4).

Figure 2.23▶
Intake and release of products from a cell

Substances taken in: Substances released:

oxygen ⟶ ⟶ carbon dioxide

glucose ⟶ ⟶ water

This process transfers a vast amount of energy. The word equation for respiration is:

$$\text{oxygen} + \text{glucose} \rightarrow \text{carbon dioxide} + \text{water} [+\text{energy}]$$

The energy transferred during respiration is used:

- to help cells build up large molecules from smaller ones, e.g. proteins from amino acids

- to help muscles contract

- to help maintain a steady body temperature in cold environments

- in the active transport of materials across boundaries.

Where does the energy come from?

Air is breathed into the lungs and oxygen diffuses into the bloodstream.

Haemoglobin in red blood cells combines with oxygen to make oxyhaemoglobin.

Food is digested and glucose is absorbed into the bloodstream.

The heart pumps blood to the tissues.

The blood contains oxyhaemoglobin in the red blood cells and glucose in the blood plasma.

Figure 2.24 ▲
Increased heart rate and breathing rate ensure that more oxygen and glucose are delivered to the muscles for running

Figure 2.25 ▼
Micrograph of a mitochondrion

What comes out?

Muscles respire producing heat, carbon dioxide and water as waste products.

The skin releases excess heat.

The heart pumps the blood containing carbon dioxide, dissolved in the blood plasma and in the red cells, from the muscles to the lungs.

Carbon dioxide is breathed out from the lungs (heat energy and water vapour are also released from the lungs).

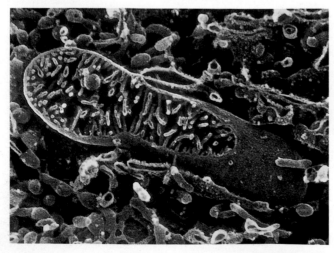

Aerobic respiration

Remember, it is called aerobic respiration because oxygen is being used.

Plants and animals need to respire all the time so that they can stay alive. We even need energy when we sleep so that our heart can pump, we can breathe and our brain can function.

Aerobic respiration takes place in the mitochondria.

Anaerobic respiration

When there is insufficient oxygen available to meet the oxygen requirements of cells they can respire without oxygen!

If one arm is raised and the other kept by your side and both fists are clenched regularly for several minutes the raised arm will start to ache quite badly.

But why does the raised arm ache so much? The arm muscles had to respire to release energy, but the raised arm did not get enough oxygen and had to respire without oxygen – **anaerobic respiration**.

Anaerobic respiration in humans

Compared to aerobic respiration, very little energy is transferred during anaerobic respiration because the breakdown of glucose is incomplete.

$$\text{glucose} \rightarrow \textbf{lactic acid} \ [+ \text{energy}]$$

The lactic acid makes the muscle hurt – it is a mild poison that causes muscle fatigue.

A sprinter can run very fast for a short period of time. The muscles respire anaerobically for a short period of time, but the lactic acid has to be removed and oxygen is used to do this. It is said that the muscles have built up an **oxygen debt** because the cells are 'owed' oxygen. That is why sprinters breathe deeper and faster than normal after the race has finished.

Much less energy is transferred by anaerobic respiration since glucose is only partially broken down.

Rapid breathing enables more oxygen to reach the lactic acid and oxidise it to carbon dioxide and water, or convert lactic acid back to glucose.

Summary

- **Aerobic respiration** can be summarised as
  glucose + oxygen $\longrightarrow$
  carbon dioxide + water [+ energy]

- The energy transferred during respiration is used to:
  - build up larger molecules from smaller ones
  - enable muscles to contract
  - maintain a steady body temperature in cold surroundings
  - enable active transport to take place.

- Aerobic respiration in cells takes place in the mitochondria.

- During vigorous exercise muscles may get short of oxygen. They then obtain energy from glucose by **anaerobic respiration**.

- **Lactic acid** is the waste product of anaerobic respiration. Lactic acid is a poison.

- Less energy is transferred during anaerobic respiration compared to aerobic respiration. This is because during anaerobic respiration the glucose is only partially broken down into lactic acid.

- Anaerobic respiration results in an **oxygen debt**.
  The oxygen is used to oxidise the lactic acid into carbon dioxide and water.

Topic questions

1   Arrange the following into the equation for aerobic respiration:

| glucose | | → | | carbon dioxide | | water | | + |

| + | | oxygen | | [+ energy] |

2   Which substance is produced during aerobic respiration by animal and plant cells?

    A Alcohol   **B** Carbon dioxide
    C Lactic acid   **D** Nitrogen

3   Cells which line the small intestine need a lot of energy to absorb digested food. Which of the following must be present in large numbers in these cells?

    A Chromosomes   **B** Fat molecules
    C Mitochondria   **D** Villi

4   Explain why it is not possible for a person to run and respire anaerobically for a long time.

5   What is meant by an oxygen debt?

Examination questions

1   Blood contains plasma, platelets, red cells and white cells. Each has one or more important functions.

Copy the table below and draw a line from each part to its function.

One part has two functions. Draw lines from this part to both functions.

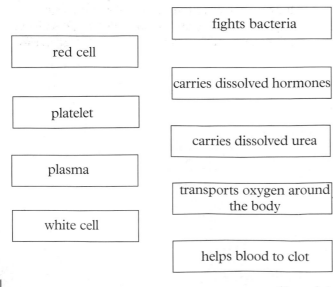

Name of part of blood

red cell

platelet

plasma

white cell

Function of part of blood

fights bacteria

carries dissolved hormones

carries dissolved urea

transports oxygen around the body

helps blood to clot

(5 marks)

2   a) The sentences are about breathing. Choose words from the list in the box to complete the sentences that follow. Each word may be used once or not at all.

| decreases | diaphragm | in | increases | lungs |
| | out | rib | ventilation | |

Air is drawn into the _____ as the thorax _____ in volume.
This change in volume is caused by the _____ muscles contracting and moving the rib cage _____ at the same time as the _____ is moved down by its muscles.

(5 marks)

3   The diagram shows a human heart.

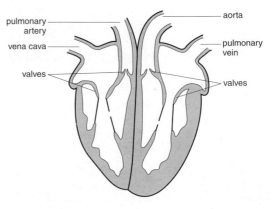

pulmonary artery

vena cava

valves

aorta

pulmonary vein

valves

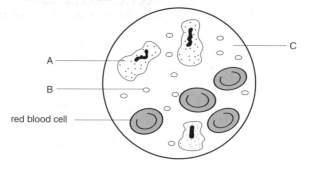

a) Complete these sentences. Use the information in the diagram to help.

Blood from the body enters the right side of the heart through the blood vessel called the

_____.

The blood is pumped through the pulmonary _____ to the lungs.
The blood returns to the heart through the blood vessel called the _____.
The blood is pumped to the body through the blood vessel called the _____. *(4 marks)*

b) What do valves do in the heart? *(1 mark)*
c) Describe the differences between arteries and veins. *(4 marks)*
d) Complete this sentence.

In the lungs the blood loses _____ gas and picks up _____ gas. *(2 marks)*

4 a) Copy and complete the table to give one site where digestive substances are made.

Digestive substance	One site of production
bile	
amylase	
lipase	
protease	

(4 marks)

b) Describe **two** ways that the mouth can break down starchy foods. *(2 marks)*
c) Describe how the small intestine is adapted for absorbing food. *(5 marks)*
d) Describe how the liver helps to digest fats. *(2 marks)*

5 The diagram shows four parts of blood.

a) Complete the table to give the name and function of the parts labelled A, B and C.

Letter	Name	Function
A	....	....
B	....	....
C	....	....

(6 marks)

b) Red blood cells contain haemoglobin. Explain how this enables red blood cells to pick up oxygen from the alveoli and release it to cells in other parts of the body. *(4 marks)*

Chapter 3
Response, co-ordination and health

Key terms

addiction · ADH · antibodies · antitoxins · bacteria · bladder · brain · cancer · ciliary muscles · constrict · cornea · diabetes · dialysis · dilate · effectors · endocrine · environment · excretion · eye · fertility drugs · filtration · focus · FSH · gland · glucagon · glycogen · homeostasis · hormones · image · insulin · iris · kidney · lens · LH · motor neurone · mucus · negative feedback · nerves · nerve impulse · neurone · nicotine · oestrogen · optic nerve · oral contraceptives · pancreas · pituitary gland · pupil · reabsorption · receptors · reflex action · reflex arc · relay (connector) neurone · renal vein · renal artery · response · retina · sclera · sensory neurone · skin · stimulus · suspensory ligaments · synapse · target organ · toxin · urea · urine · vaccines · virus

3.1

Co-ordinated	Modular
10.8	Mod 02
	11.5

Nervous system

Stimulus and response

Living things need to be able to respond to the environmental changes that go on around them for many reasons. A change in the **environment** that affects living things is called a **stimulus** and the reaction by the animal or plant is called a **response**. In humans most responses to a stimulus are brought about by electrical impulses which pass along nerves. There are two important types of responses:

- voluntary responses. These require thought and thus involve the **brain**. Speech is an example of a voluntary response.

- automatic responses called **reflex actions**. These do not involve the brain directly. The knee jerk reflex is an example.

The nervous system

Nerves are made up of vast numbers of nerve cells called **neurones**. The structures we call nerves within the body are made up of bundles of these neurones.

There are several types of neurones. **Sensory neurones** receive information from **receptors** that are found in all parts of the body. Receptors are cells which can detect stimuli. They may be scattered all over the body, such as touch receptors, or they may be grouped into sense organs.

Each receptor is sensitive to one type of stimulus only, so we have receptor cells that are sensitive to:

- changes in position; these help with balance, and are found in the ear
- light, found in the eye
- chemicals, found in the nose; these enable us to smell
- chemicals, found in the tongue; these enable us to taste
- sound found in the ear
- touch, pressure and temperature, all of which are found in the skin.

Effectors carry out responses. They are usually muscles or **glands**. **Motor neurones** carry the information which tells the effectors how to respond.

Figure 3.1
The human
nervous system

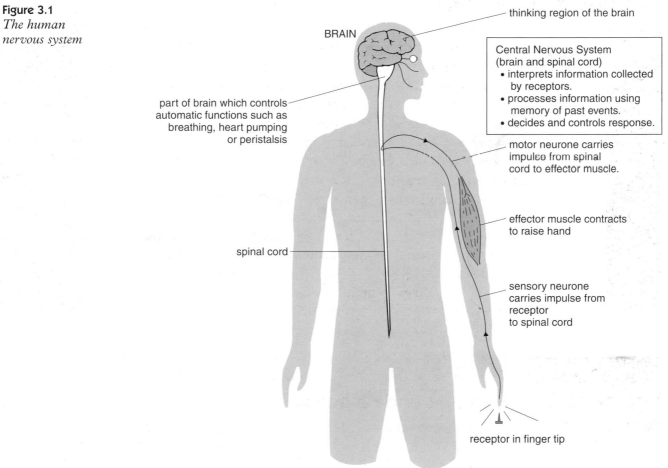

thinking region of the brain

BRAIN

Central Nervous System
(brain and spinal cord)
- interprets information collected by receptors.
- processes information using memory of past events.
- decides and controls response.

part of brain which controls automatic functions such as breathing, heart pumping or peristalsis

motor neurone carries impulse from spinal cord to effector muscle.

effector muscle contracts to raise hand

spinal cord

sensory neurone carries impulse from receptor to spinal cord

receptor in finger tip

Reflex actions

Examples of reflex actions include:

1 the knee jerk reflex – the purpose of which is to maintain the posture of the body

2 removing the hand or fingers from a hot object – the purpose of which is to protect the skin from burning

3 the dilation and constriction of the **pupil** in the eye to protect the **retina** from damage by bright light.

Reflex actions usually

- protect the body from harm.

- follow a fixed pathway – **nerve impulses** are sent by receptors through the nervous system to effectors. This pathway is called a **reflex arc**.

Did you know?

Because many of the effectors are muscles, the term 'motor neurone' is often used instead of the correct term 'effector neurone'.

The reflex arc

In a simple reflex action electrical impulses are carried from receptors, such as the pressure receptors in your hand, along sensory neurones to the spinal cord. The spinal cord contains **relay neurones** (sometimes called **connector neurones**). These link the sensory neurones with the **motor neurones**. These neurones carry the electrical impulses to the **effectors**. It is the effectors that carry out the responses. Effectors can be muscles or glands.

Where the ends of the neurones come close together in the spinal cord are small gaps called synapses. When an electrical impulse reaches the end of the sensory neurone a chemical change takes place in the synapse. The chemicals released into the gap between the neurones cause the connector neurone to conduct the electrical impulse. A similar action takes place between the ends of the connector neurone and the effector neurone so allowing a continuous flow of impulses to occur between the receptor and the effector through the reflex arc. If the effector is a muscle then it contracts, if the effector is a gland then it secretes a chemical.

A reflex arc is shown in Figure 3.2.

Figure 3.2
Section through the spinal cord showing a reflex arc from the hand

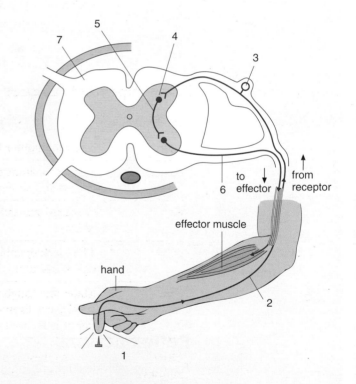

Key

1 Pressure receptor
2 Sensory neurone
3 Sensory neurone cell body
4 Synapse between sensory neurone and relay neurone
5 Relay neurone
6 Motor neurone
7 Spinal cord

In this reflex action,

- the pin pressing on the skin is the stimulus,
- the pressure receptor detects the pressure of the pin,
- the spinal cord is the co-ordinator,
- the muscle is the effector,
- pulling the arm away is the response.

The eye and light

Receptors in the retina at the back of the eye change light into electrical impulses. These impulses are carried along the **optic nerve** to the brain. The brain then changes the impulses into pictures.

Figure 3.3
A vertical section through a human eye

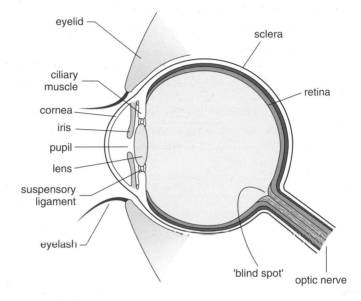

Some important features of parts of an eye

Part	Some important features
sclera	The tough outer layer of the eye to which eye muscles are attached.
cornea	The transparent part of the sclera that allows light into the eye. It is curved and so helps the light to bend inwards on its pathway to the lens.
iris	Coloured muscular layer that controls the amount of light reaching the retina.
pupil	The black opening in the middle of the iris through which light passes on its way to the lens.
lens	Made of a tough transparent jelly, held in place by ciliary muscles and suspensory ligaments. The lens causes the light to bend even more so that it is focused on the retina.
ciliary muscles	• As these muscles contract they squash the **suspensory ligaments** making them go slack and forcing the lens to become fatter so helping to focus light from a nearby object. • As these muscles relax the suspensory ligaments become stretched. This makes the lens become thinner, so helping to focus light from a distant object.
retina	It is on this layer that the light is focused. It contains receptor cells that are sensitive to light.
optic nerve	Links the retina with the brain. The receptor cells in the retina send electrical impulses to the brain along the sensory neurones in the optic nerve.

What happens when light enters the eye

Parallel rays of light enter the eye through the transparent window at the front called the **cornea**. The surface of the cornea is curved and this bends or refracts the light rays helping to **focus** them. The light rays then pass through the **lens**. The surface of the lens is also curved and the light rays are bent even more. The **ciliary muscles** can alter the curve of the lens. This helps the eye to focus on either near or distant objects (Figure 3.4).

Figure 3.4
The shape of the lens changes in order to focus either close up or far away

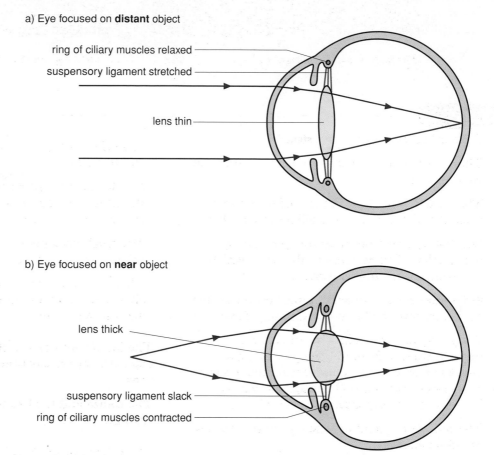

a) Eye focused on **distant** object

ring of ciliary muscles relaxed
suspensory ligament stretched
lens thin

b) Eye focused on **near** object

lens thick
suspensory ligament slack
ring of ciliary muscles contracted

The light rays focused by the cornea and the lens fall on to the **retina** where they form a sharp **image**. The retina is very sensitive to light and the cells change the light into nerve impulses that are carried to the brain along the optic nerve.

The **iris** controls the amount of light that reaches the retina. The iris is a ring of muscle in front of the lens. When the circular muscles in the iris are contracted, the gap or pupil through which the light can pass is very small and only a little light reaches the retina. When the circular muscles are relaxed, the pupil is large. The light is controlled in this way to protect the very delicate light receptor cells of the retina from damage by bright light.

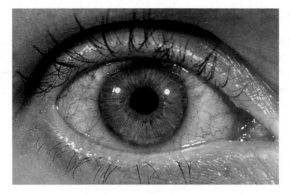

The pupil is small in bright light . . .

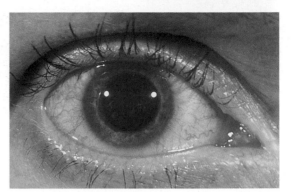

. . . but larger in dim light

Summary

- Cells called **receptors** detect **stimuli**.

- These stimuli are any changes that occur in the environment.

- The human body has a variety of receptors, each of which can detect a different stimulus.

- **Electrical impulses** from receptors pass along nerve cells called **neurones** to the brain which co-ordinates any **response** needed.

- **Reflex actions** are automatic, rapid responses that do not involve the brain coordinating them.

- Simple reflex actions involve an electrical impulse passing from the receptor along a **sensory neurone** to the spinal cord, then along a **motor neurone** to a muscle or a gland which brings about a response.

- The muscle or gland producing the response is called an **effector**.

- The spinal cord contains relay (connector) neurones which link the sensory neurones with the motor neurones during a reflex action.

- In a reflex action the spinal cord acts as the co-ordinator.

- A reflex action can be summarised as:
  stimulus → receptor → co-ordinator → effector → response

- The small gaps between the neurones in the spinal cord are called synapses.

- Chemicals are released at each synapse which allow the electrical impulses to pass from one neurone to the next.

- The important parts of the eye are the **sclera, iris, pupil, lens, suspensory ligament, retina** and **optic nerve.**

- The **cornea** and the **lens** together focus light onto the retina.

- The shape of the lens is controlled by the ciliary muscles.

Topic questions

1 Where in our bodies are receptors which are sensitive to:
   a) light
   b) sound
   c) changes in body position
   d) chemicals
   e) touch
   f) pressure
   g) temperature changes?

2 What travels along neurones?

3 What is the name of the muscles or glands that carry out responses?

4 a) What is a reflex action?

   b) Put the following in the order electrical impulses flow in a reflex action.

   **effector   motor neurone   receptor
   sensory neurone   spinal cord**

5 Copy out then complete the gaps in the following table.

Part	Some important features
	This is the tough outer layer of the eye to which eye muscles are attached.
	This is transparent and allows light into the eye.
	This is curved and so helps the light to bend inwards on its pathway to the lens.
iris	
	This is the black opening in the middle of the iris through which light passes on its way to the lens.
	This is made of a tough transparent material and causes the light to bend so that it is focused on the retina.
retina	
	This links the retina with the brain.

6 Describe how the action of the ciliary muscles helps the lens to focus on a distant object.

7 a) What is a synapse?
   b) How are electrical impulses passed across a synapse?

8 Put the following in the order the electrical impulses flows in a reflex action.

relay neurone   effector   motor neurone
receptor   response   sensory neurone   stimulus

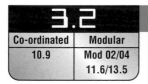

Hormones and the menstrual cycle

Co-ordinated	Modular
10.9	Mod 02/04
	11.6/13.5

Not all the information within the human body is transmitted through nerves. It is also transmitted by **hormones**. Hormones are chemical substances secreted directly into the blood plasma by special glands called **endocrine glands**. They are carried in the blood plasma to a **target organ** in another part of the body. Hormones often control our long-term responses to changes in the body or the environment.

Sex hormones

The monthly release of an egg from a woman's ovaries and the changes in the thickness of the lining of her womb are controlled by hormones secreted by her **pituitary gland** and ovaries. The stages in this menstrual cycle are shown in Figure 3.5

The hormones involved include oestrogen, secreted by the ovaries, follicle-stimulating hormone (FSH) and luteinising hormone (LH) which are both secreted by the pituitary gland.

Hormones and fertility

Artificial sex hormones are now manufactured and given to some females who do not produce enough of their own or who are infertile. They help the ovaries to make more eggs than usual, increasing the chances of one of these eggs being fertilised by a sperm from a male. This increases the fertility of the female and are called fertility drugs.

Figure 3.5
Hormones and the menstrual cycle

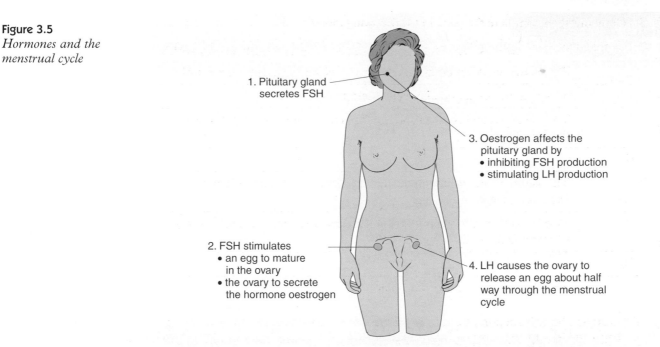

1. Pituitary gland secretes FSH

3. Oestrogen affects the pituitary gland by
   • inhibiting FSH production
   • stimulating LH production

2. FSH stimulates
   • an egg to mature in the ovary
   • the ovary to secrete the hormone oestrogen

4. LH causes the ovary to release an egg about half way through the menstrual cycle

Some women do not produce enough FSH to stimulate eggs in the ovary to mature. Doctors may give these women FSH. FSH acts as a fertility drug by stimulating eggs to mature. If the FSH dose is too high, several eggs may be released at the same time. If these eggs are fertilised, several babies will develop in the woman's womb. This is called a multiple pregnancy. Often few of these babies survive.

Other hormones can be given that prevent the release of eggs from the ovaries. Such hormones are **oral contraceptives** or birth control pills.

The birth control pill contains hormones, including oestrogen that inhibit FSH production. Without FSH no eggs mature, therefore the woman cannot become pregnant.

The use of oral contraceptives rather than barrier methods (condoms) may result in the spread of sexually transmitted diseases.

Summary

◆ **Hormones** are chemicals that control many processes within the body. They are secreted by **glands** and are transported by the bloodstream to **target organs**.

◆ In women, hormones secreted by the **pituitary gland** and the ovaries control the monthly release of an egg and the changes in thickness of the lining of the womb.

◆ FSH (follicle stimulating hormone), secreted by the pituitary gland, causes an egg in the ovaries to mature and this stimulates the ovaries to produce hormones including oestrogen.

◆ High levels of oestrogen stop the further production of FSH, but stimulate the production of LH.

◆ LH (luteinising hormone) secreted by the pituitary gland, stimulates the release of an egg about half way through the menstrual cycle.

◆ **Fertility drugs** are hormones that stimulate the release of eggs from the ovary.

◆ **Oral contraceptives** are hormones that prevent the release of eggs from the ovaries.

◆ There are benefits and problems regarding the use of hormones to control fertility.

Topic questions

1 Complete the gaps in the following sentence. Hormones are secreted by _____ and are transported to their _____ organs by the _____.

2 Where are the hormones that control the monthly release of an egg from a women's ovary and the changes in thickness of the lining of her womb produced?

3 a) What is the action of a fertility drug?
  b) What is the action of an oral contraceptive?

4 A couple are finding it difficult to conceive. After tests have been carried out, the man is found to have fewer live sperm than normal. The woman is given an artificial hormone to increase the number of eggs she produces. Explain why this may help her to conceive even though the man has a low sperm count.

5 a) What do the initials FSH mean?
  b) Which gland secretes FSH?
  c) Describe two functions carried out by FSH.

6 a) What do the initials LH mean?
  b) Which gland secretes LH?
  c) What function is carried out by LH?

7 What effect does oestrogen have on:
  a) the action of FSH?
  b) the action of LH?

3.3 Hormones and diabetes

Co-ordinated	Modular
10.9	Mod 02
	11.6

Controlling blood sugar

The glucose concentration of the blood is both monitored and controlled by the pancreas. Blood sugar concentration is controlled by two hormones which are secreted by the **pancreas** – **insulin** and **glucagon**.

A rise in blood glucose concentration stimulates the cells in the pancreas to secrete insulin into the blood. One effect of insulin is to cause the liver to convert excess glucose into insoluble **glycogen** and store it.

A fall in blood glucose concentration stimulates other cells in the pancreas to secrete the hormone glucagon into the blood. This causes the liver to convert glycogen into glucose and release it into the blood.

Unfortunately there are some people who cannot produce insulin either at all or in sufficient quantities to control blood glucose. These people suffer from **diabetes**. Mild cases of diabetes are treated by carefully controlling fat and carbohydrate intake or by injecting insulin into the blood.

Figure 3.6
Feedback system for the control of blood glucose concentration

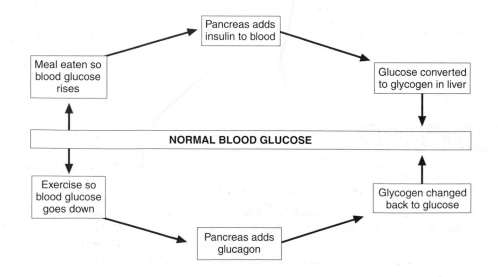

Did you know?

In diabetes:

- the liver cannot convert glucose to glycogen and the concentration of glucose in the blood rises to very high levels.

- the first sign of a problem is when glucose begins to appear in the urine because the kidney cannot reabsorb all of it from the urine.

- the body cells are also starved of glucose because insulin is needed for them to be able to use it.

- most cells can use alternative sources of sugar but the brain's cells can only use glucose. Prolonged deficiency of glucose by brain cells in diabetes leads to blackouts and coma (and if not treated – death).

- the cells of the retina in the eye may also be affected, leading to blindness.

- the diet has to be carefully controlled because sudden intakes of large amounts of carbohydrates cause the blood sugar level to rise and stay high (diabetics often develop glucose tolerance).

- the most suitable treatment, if control by diet is insufficient, is regular injections of insulin.

- the injections of insulin can also be a problem because a sudden injection of a large quantity of insulin could cause the glucose level in the blood to fall too fast to below the safe level. For this reason, diabetics may carry sugar in the form of sweets or sugar lumps to give the glucose levels a quick boost if they fell dizzy or faint after an insulin injection.

Did you know?

Normally there are about 7 g of glucose circulating in the blood at any one time and this is sufficient for the normal working of the body. However, there are times when blood glucose level rises or when more glucose needs to be available for the production of extra energy.

The level will rise immediately after a meal, and more glucose will be needed in the blood for the production of energy during vigorous activity.

Summary

- Blood glucose concentration is controlled by **hormones** called **insulin** and **glucagon**. These are secreted by the **pancreas**.

- The pancreas monitors and controls the blood glucose concentration.

- If the blood glucose concentration is too high the pancreas secretes insulin. This causes the liver to convert glucose into insoluble glycogen and store it.

- If the blood glucose concentration is too low the pancreas secretes glucagon. This causes the liver to convert glycogen into glucose which is released into the blood.

- **Diabetes** is caused when a person's pancreas does not secrete enough insulin.

- Diabetes can be treated by a carefully controlled diet or by insulin injections.

Topic questions

1 a) Which two hormones control the concentration of glucose in the blood?
  b) Where are these two hormones produced?

2 What happens to the concentration of blood glucose if not enough insulin is produced?

3 a) What happens in the pancreas and the liver if the blood concentration becomes too high?
  b) What happens in the pancreas and the liver if the blood glucose concentration becomes too low?

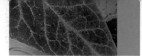

In order to remain healthy, the conditions inside our bodies (the internal environment) must be kept the same. If the conditions are not kept the same, our body systems will not work properly. The processes through which the body monitors and controls its internal environment are called **homeostasis**.

The body produces several waste products including:

* carbon dioxide from respiration, which leaves the body via the lungs when we breathe out,

* urea which is produced in the liver by the breakdown of excess amino acids. Urea is removed by the kidneys in urine.

It is important that the body is able to monitor and control the amounts of these waste products. If they build up they will poison the body.

Four conditions that need to be kept the same are:

* the amount of water in the blood
* the temperature of the body
* the amount of glucose in the blood (see section 3.3).
* the ion content of the blood.

Controlling the amount of water in the blood

Each day different volumes of liquids are drunk, different amounts of exercise are taken, different volumes of sweat are produced and different volumes of water are breathed out and yet the water in the blood remains at a constant concentration.

Figure 3.7
Daily water inputs and outputs

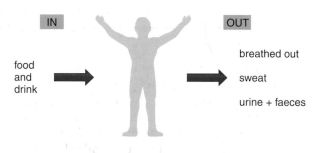

The **kidneys** play an important part in keeping the water concentration in the blood constant. The kidneys are part of the excretory system.

Figure 3.8
The excretory system in humans

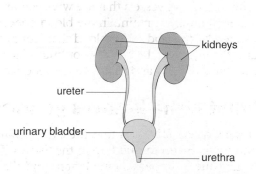

The excretory system

Arterial blood arrives at the kidney under high pressure. Blood leaves the kidney through the vein, joining the vena cava on the journey back to the heart. The venous blood will not only have lost pressure but will also have lost any excess water, ions and urea in the kidney.

Figure 3.9
The blood supply to the kidneys

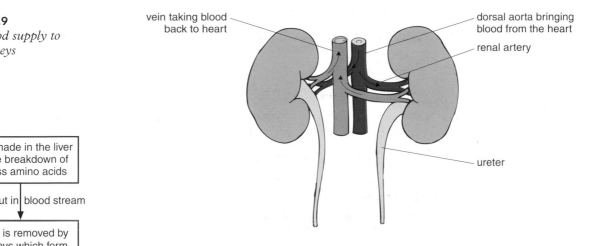

vein taking blood back to heart

dorsal aorta bringing blood from the heart

renal artery

ureter

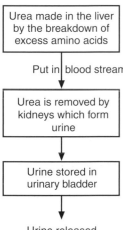

Urea made in the liver by the breakdown of excess amino acids

Put in blood stream

Urea is removed by kidneys which form urine

Urine stored in urinary bladder

Urine released through urethra

Figure 3.10
The process of removal of urea

The kidney makes **urine** with the excess water, excess ions and **urea**. Adults make about 1.7 litres of urine a day, depending on fluid intake, sweat etc. The urine is stored in the **bladder** before being excreted.

What happens in the kidneys?

As blood is circulated round the body it passes through the kidneys where it is filtered, the urea is removed, turned into urine and the concentration of urine is controlled.

Each kidney has many millions of kidney tubules. Each tubule is served by a network of capillaries. As the blood flows through the capillaries, it is filtered by the tubules. As the filtrate flows through each of the tubules, the urine is gradually formed. Blood in the arteries is under high pressure. Because the artery divides suddenly into the many narrow capillaries which surround each tubule, the tubule is also under high pressure. It is this pressure which results in filtration (Figure 3.11).

The high pressure causes a large volume of liquid and the molecules that are small enough to pass through the cell membranes, to enter the tubule. The problem is that too much water and too many useful molecules may have left the blood by filtration, so there is now a system to reclaim what the body needs. The tiny blood vessels that are wrapped around the tubule selectively reabsorb the water required to maintain the blood concentration, and the glucose and ions needed by the body. The blood then continues on to the vein leaving the kidney and the liquid in the tubules continues its journey to join with the contents of all the other tubules and finally to the bladder as urine.

What if the kidneys go wrong?

If the kidneys fail to remove the excess water and ions, the body retains the fluid and it will begin to build up in the tissues. This puts a considerable strain on the heart (due to impaired circulation) and the person can become very ill.

Figure 3.11
Filtration by a tubule

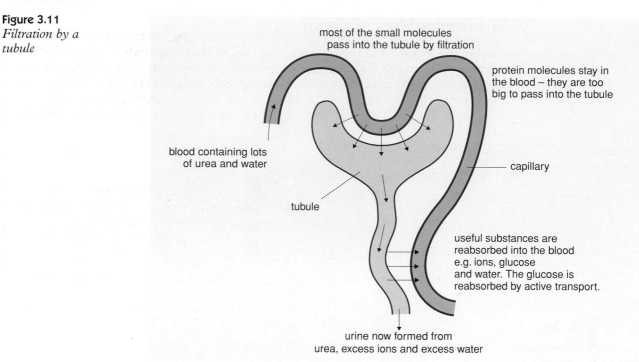

most of the small molecules pass into the tubule by filtration

protein molecules stay in the blood – they are too big to pass into the tubule

blood containing lots of urea and water

capillary

tubule

useful substances are reabsorbed into the blood e.g. ions, glucose and water. The glucose is reabsorbed by active transport.

urine now formed from urea, excess ions and excess water

Did you know?

Kidneys and dialysis

One way of treating the failure to remove water and salts is to take the blood from a vein and pass it through a system that will do the job of the kidneys, by removing the water and salts. The blood is then put back into the vein and continues its journey around the body. One problem with this system is that it can only work while the person is linked to the 'dialysis' machine; the rest of the time the blood and body fluids continually move away from the right concentration. To minimise these effects, patients with failing kidneys are very careful with their diets and try to ensure that they do not put too great a strain on their body systems. However dialysis still has to take place regularly and is very time consuming.

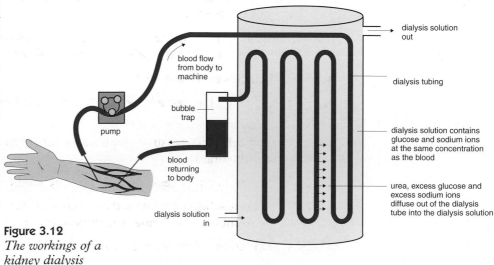

dialysis solution out

blood flow from body to machine

dialysis tubing

bubble trap

pump

dialysis solution contains glucose and sodium ions at the same concentration as the blood

blood returning to body

urea, excess glucose and excess sodium ions diffuse out of the dialysis tube into the dialysis solution

dialysis solution in

Figure 3.12
The workings of a kidney dialysis machine

The blood and dialysis fluids are separated in the dialysis machine by fine layers of dialysis membrane (a bit like 'Clingfilm'). The blood is under quite high pressure flowing through the machine. The dialysis fluid is at the same concentration as the blood plasma (including sodium ions and glucose). The fluid does not have urea or excess water – both of which are present in the blood – this means that the urea, excess water and sodium ions diffuse out of the blood into the dialysis fluid and are thereby removed from the body. The 'cleaned' blood can then be returned to the patient.

Kidney transplants

If, as a result of kidney failure, a person needs dialysis regularly, one alternative is to have a kidney transplant. The donated kidney can come from a living person – generally a relative – or from an anonymous donor who has recently died. The kidney has to be a good blood and tissue match or the body's immune system will reject it. Even so, drugs are given to help the body accept the transplanted kidney.

Once the kidney is in place and hasn't been rejected by the body, the person can once again live a normal life with no need for dialysis.

How the water content of the blood is controlled

In order to keep the internal environment of the body stable, a number of automatic control systems are at work. The self-regulation requires the system to be continuously monitored. For example, if the body gains too much water then the system tries to restore the original balanced conditions by losing water. This automatic response is called negative feedback. Negative feedback is the system by which the desired factors are kept constant – it means that when one of the constants is changed, a response resulting in the opposite effect occurs.

Water concentration in the blood is controlled by the reabsorption of water by the kidneys. As blood flows through the blood vessels at the base of the pituitary gland in the brain, specialised cells measure the water content of the blood. If the blood is getting short of water because the person has not had a lot to drink or has lost a lot of water by exercise, then:

- the pituitary gland releases a hormone into the blood. The hormone is known as ADH (anti-diuretic hormone)

- when the blood containing the ADH reaches the kidneys, the cells of the kidney tubules let more water be reabsorbed into the capillaries.

- the urine then contains less water.

If, however, the person has had a lot to drink or has not lost a lot of fluid via exercise, then:

- the specialised cells will measure a high water concentration in the blood as it flows through the brain

- the pituitary gland stops releasing ADH into the blood

- when the blood reaches the kidneys the absence of ADH causes the tubules to reduce the amount of water being reabsorbed into the blood

- a large volume of urine is then produced, containing a lot of water.

51

Figure 3.13
The processes that occur when the blood contains a) too little and b) too much water

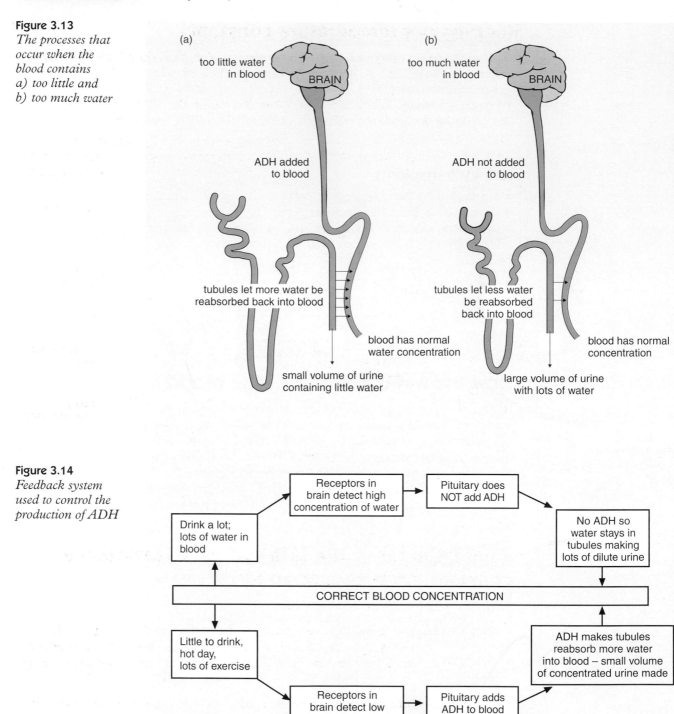

Figure 3.14
Feedback system used to control the production of ADH

Keeping cool and keeping warm – the skin and its blood vessels

The **skin** has three important functions:

- it protects the body from invasion by bacteria, viruses and fungi.
- it is waterproof as a result of the outer layer.
- it helps to maintain a constant body temperature of about 37°C.

Keeping our temperature constant

It is vital to humans and all warm-blooded animals that a constant body temperature is maintained to enable all the enzymes to work efficiently.

Body temperature is monitored and controlled by the thermoregulation centre in the brain. Temperature sensors in the body monitor the temperature of the blood.

If the blood is too hot or too cold the body has a system for keeping cool and another for keeping warm. Blood flows from the arteries towards the surface of the skin in tiny vessels that further divide into smaller vessels, eventually recombining to become veins.

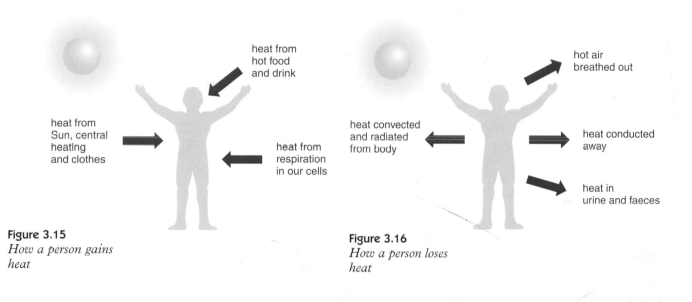

Figure 3.15
How a person gains heat

Figure 3.16
How a person loses heat

What happens at the skin surface to maintain a constant internal temperature?

When our body starts to get too hot:

- blood vessels supplying the capillaries near the surface of the skin get wider (dilate). This allows more blood to flow through the capillaries near the surface of the skin causing it to redden. Heat from this blood is transferred to the air by radiation.

- more sweat is produced by the sweat glands. The water in sweat evaporates. In order to evaporate, the water needs heat energy. The heat energy needed to cause the evaporation comes from the body, so the body cools down.

When our body starts to get too cool:

- blood vessels supplying the tiny capillaries get narrower (constrict), reducing the blood flow to the capillaries near the surface of the skin. This causes the skin to go white and limits the heat loss by radiation.

- almost no sweat is produced, reducing heat loss further. The reduction of blood flow towards the surface of the skin ensures that the blood for the vital organs is at the correct temperature.

At no time either in keeping cool or staying warm do the blood vessels actually move!

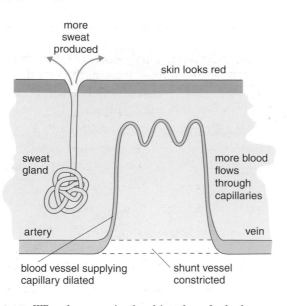

Figure 3.17 *What happens in the skin when the body gets too hot*

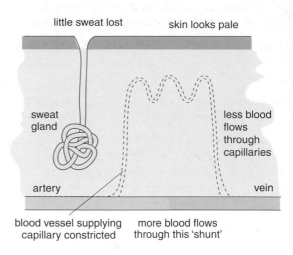

Figure 3.18 *What happens in the skin when the body gets too cool*

Shivering

When the core body temperature is too low, the body may shiver. Shivering is the involuntary contraction of several of the muscles. Muscle contraction uses energy from respiration. Some energy from respiration in muscles is released as heat so blood flowing through these muscles is warmed.

Therefore thermoregulation of the skin keeps the body temperature constant.

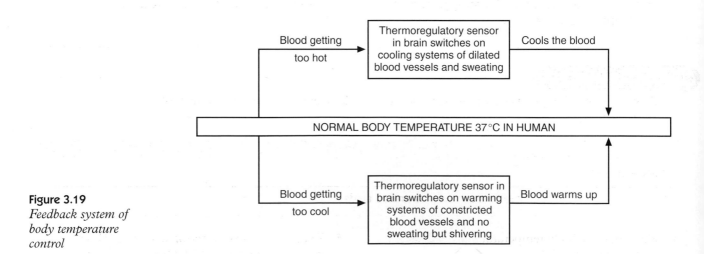

Figure 3.19
Feedback system of body temperature control

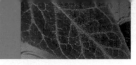

Summary

- Carbon dioxide produced by respiration is removed via the lungs during breathing out.

- **Urea** produced in the liver by the breakdown of excess amino acids is removed via the kidneys in the **urine**.

- **Homeostasis** is the process by which the internal conditions are self-regulated.

- The water content of the body needs to be kept constant. Water is lost via the skin as sweat, via the lungs and via the kidneys in urine. To balance these losses more water needs to be taken in as food and drink.

- The ion content needs to be kept constant. Ions are lost via the skin as sweat, and via the kidneys in urine.

- The kidneys maintain the internal environment by:

  - **filtering** the blood

  - **reabsorbing** all the glucose

  - reabsorbing the amount of dissolved ions needed by the body

  - reabsorbing as much water as the body needs

  - releasing urea, excess ions and excess water as urine.

- The internal temperature needs to be kept constant. This maintains the temperature at which enzymes work best.

- The thermoregulatory centre in the brain monitors the temperature of the blood. If the blood temperature is too high:

  - the blood vessels supplying the skin capillaries dilate so more blood flows through the capillaries so losing heat

  - the sweat glands release more sweat which cools the skin as it evaporates.

  If the blood temperature is too low:

  - the blood vessels supplying the skin capillaries constrict so less blood flows through the capillaries so losing less heat

  - the muscles 'shiver' because the muscles repeatedly contract and relax. Each contraction needs the energy transferred during respiration. Some of this energy is transferred as heat.

- Receptors in the brain monitor the concentration of water in the blood. If there is too little water in the blood:

  - the pituitary gland secretes a hormone called ADH (anti-diuretic hormone) into the blood.

  - the kidneys reabsorb more water, resulting in urine being more concentrated.

  If there is too much water in the blood:

  - less ADH is secreted and the concentration of water in the blood increases.

Topic questions

1 Why do we need to get waste products removed from our body?

2 Complete the gaps in the table.

Waste product	How produced	How removed
Carbon dioxide		
	breakdown of excess amino acids	

3 Why does our body temperature have to remain constant at about 37°C?

4 a) Where is urea made?
   b) What is urea made from?
   c) Where is urine made?
   d) What is urine made from?

5 Explain, using your knowledge of ADH, why on a hot day when James had been running in a cross-country race he did not produce much urine and was very thirsty.

6 What does the body do to keep the internal organs warm on a cold day? Include as much detail as possible.

3.5

Co-ordinated	Modular
10.11	Mod 01
	10.5

Fighting disease

Bacteria and viruses

Most diseases are caused by two types of microorganisms, **bacteria** and **viruses**.

Although there are many different types of bacteria, the cells of each contains:

- cytoplasm
- a cell membrane
- a cell wall
- genetic material

Bacterial cells do not have a recognisable nucleus.

Diseases caused by bacteria include tetanus, diarrhoea, cholera and food poisoning.

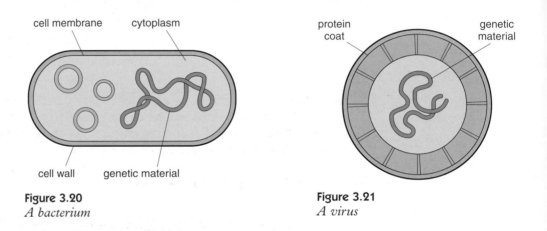

Figure 3.20
A bacterium

Figure 3.21
A virus

Viruses are very much smaller than bacteria.

The cell of each virus consists only of a protein coat which surrounds a few genes.

Viruses are unaffected by antibiotics. Because they reproduce only in living cells they are difficult to attack without destroying the living cells as well.

Different viruses attack different cells.

- The virus that causes the common cold and 'flu' attacks the cells lining the respiratory system.

- The AIDS virus attacks the blood cells which help our bodies fight disease.

- The rubella (German measles) virus can seriously damage the nervous system of a fetus during the first few months of its development.

The body's defences against disease

The body has several methods of defending itself against the entry of microorganisms.

- The skin acts as a barrier to prevent the entry of disease-causing microorganisms. Any cuts should be cleaned and covered with a plaster or bandage. This covering stops the microorganisms from getting into the bloodstream and speeds up the healing process.

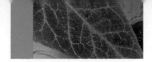

- The trachea and the air passages in the lungs are lined with special cells that have hair like structures called cilia. In between these cells are other types of cells that produce a sticky liquid called **mucus**.

Figure 3.22
Cross-section of the mucus-producing cells that line the air passages of the respiratory system

flow of mucus (carrying trapped dirt and germs)

cilia

mucus producing cell

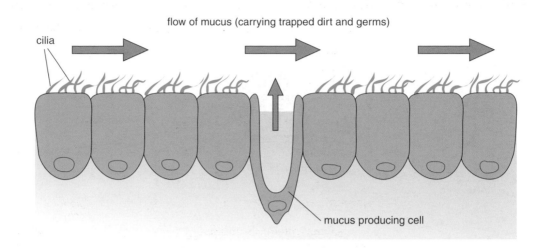

This mucus traps many of the microorganisms that enter the body every time air is breathed in. The cilia are constantly beating in a direction away from the lungs. Their beating motion moves the mucus and any microorganisms that have been trapped towards the mouth where it gets swallowed.

- The blood produces clots which seal cuts.

Figure 3.23
Fibrin threads trap the blood cells to form a clot

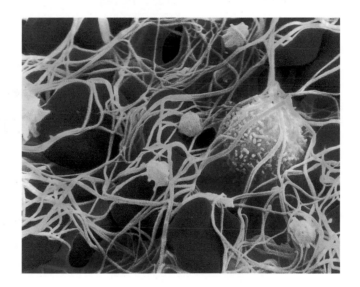

Platelets are cell fragments which play a very important role in the formation of blood clots. At a wound, platelets and cut capillaries release a chemical which starts a chain of enzyme reactions. The final reaction causes a protein in the plasma to form fine threads that then trap the platelets and cells forming a clot that eventually dries to make a scab. The clot serves two functions – it stops further bleeding and prevents entry of harmful microorganisms.

The work of white blood cells

Some white blood cells fight against infection by ingesting and destroying bacteria.

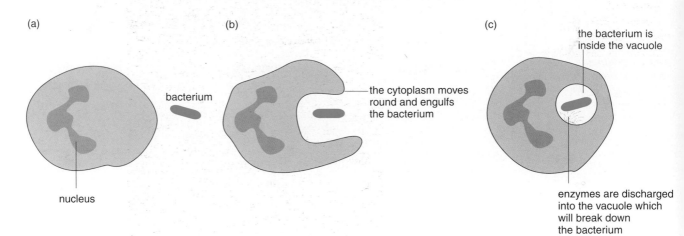

(a)

nucleus

(b)

bacterium

the cytoplasm moves round and engulfs the bacterium

(c)

the bacterium is inside the vacuole

enzymes are discharged into the vacuole which will break down the bacterium

Figure 3.24
A white cell engulfing and destroying a bacterium

Another type of white cell engages in 'chemical warfare' in our bodies. Invading bacteria have 'marker proteins' on their cell walls and our bodies recognise these as 'foreign'. White cells produce specific **antibodies** which cause the invading bacteria to stick together so that they can be attacked and destroyed by bacteria-engulfing white cells.

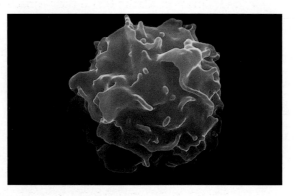

Figure 3.25
White blood cells have a very large round nucleus. these particular white cells produce antibodies

Certain invading bacteria produce chemical **toxins**. Another group of white cells produce chemicals called **antitoxins** to neutralise these toxins.

How the blood fights infection		
By attack	white blood cells	take in and destroy bacteria
By chemical action	white blood cells	• destroy bacteria by protein chemicals called antibodies. • neutralise poisons produced by bacteria by making antitoxins.
By clotting	platelets and cells are trapped by fibrin threads	this forms a clot which dries to a scab preventing • further loss of blood • entry of bacteria.

Protection by immunisation

Figure 3.26
This girl is being immunised against German measles (Rubella)

It is now possible to provide **immunity** against many diseases by introducing into a person's bloodstream dead or mild forms of the particular strain of disease-producing organism. These dead or mild forms are called **vaccines**. When these dead or mild forms enter the blood they cause some of the white blood cells to produce **antibodies** that attack and destroy the particular strain of bacterium or virus. The antibody-making cells are then ready and available to produce more antibodies very quickly should the same strain of bacterium or virus get into the bloodstream at a later date.

Did you know?

The making of a mild form of a particular strain of disease-producing organism may take many years. It usually involves the selective sub-culturing of the organism over many generations.

Flu can be caused by over 100 different strains of virus. This is why it is possible to catch flu more than once.

Did you know?

The work of Edward Jenner (1749–1823)

Edward Jenner was a doctor who worked in a country practice in Berkeley in Gloucestershire.

During the 18th century smallpox was a cause of many deaths. Even those that recovered were scarred for life by the blisters that formed on the skin.

Cowpox is a disease of cows, which shows up as small blisters containing pus on their udders and teats. Humans, such as milkmaids, who touched these parts often caught cowpox but survived. Jenner noticed that those people who had once had cowpox were not affected by smallpox. He believed that in some way having cowpox prevented people from having smallpox. In 1796 he inoculated an eight-year-old boy with fluid from a person who suffering from cowpox. Six weeks later Jenner inoculated the boy with the pus from the blisters of someone who had smallpox. The boy survived. Although there were many people who thought Jenner had taken too many risks to prove his beliefs the procedure was soon adopted and deaths due to smallpox decreased.

Both cowpox and smallpox are now known to be caused by a virus. The procedure of inoculating people with the cowpox virus in order to protect them against smallpox was the first example of **immunisation**. Smallpox has now been eradicated worldwide – no immunisation is used anywhere in the world.

Ring a ring of roses
A pocket full of posies
A tishoo! A tishoo
We all fall down

This nursery rhyme is supposed to have been linked with the Great Plague. Between 1664 and 1666 the plague killed over 10% of the population of London. The plague was carried by rats that infested the houses and streets and was passed to humans by fleas. The 'ring of roses' refers to a rash which was an early symptom of the plague. People carried posies (bunches) of flowers to hide the smells of the rat-infested streets. They thought that the smell of the flowers would stop them catching the disease. Sneezing was also a sign of having the plague. The posies did not prevent the plague and so many of them 'fell down dead'.
What they did not realise was that in order to stop the spread of such diseases there was a need to improve the way sewage was got rid off, clean up the drinking water and improve the living conditions.

At the time of the Great Plague, poor living conditions caused the rapid spread of disease. Figure 3.27 shows the conditions which still existed in slums in London in 1900. Even today people in many parts of the World live in overcrowded conditions with no adequate sewage or clean drinking water.

Figure 3.27
Slums in London during Victorian times

Some health problems of today

- Even though food-hygiene regulations have been brought into action in many countries of the world, people are still careless in the way food is grown, prepared and cooked so there are always outbreaks of food poisoning.

- Smoking cigarettes, excessive drinking of alcohol, overeating, lack of exercise and unprotected sex are recognised as lifestyles which cause disease.

- BSE (mad cow disease) was first diagnosed in cattle in Britain in 1986. In an attempt to stop the spread of BSE and variant CJD in Britain the feeding to cattle of recycled animal tissues was banned in 1988. It was not until 1996 that the eating of beef products from cattle infected with BSE was officially recognised as causing a brain disease in humans – new variant CJD (Creutzfeldt-Jakob disease).

Figure 3.28
Some milestones that have reduced the spread of disease

Date	Milestone
1790s	Edward Jenner developed the technique of immunisation
1860s	Louis Pasteur showed that food could be prevented from going bad if microorganisms from the air were kept away from the food
1865	Joseph Lister carried out the first operation in which carbolic acid was used to kill the microorganisms on the instruments and bandages. Carbolic acid was the first antiseptic to be used.
1859–1875	The first attempt to stop London's sewage going directly into the river Thames.
1928	Alexander Fleming developed the first antibiotic – penicillin.
1945	DDT was used world-wide to fight insect-borne diseases such as malaria and typhus. Its use is now been banned in many countries.
1948	National Health Service was introduced into Britain to provide free health care for all
1956	Clean Air Act was introduced which banned the use of coal for burning in fires in London. This dramatically reduced the number of deaths from respiratory diseases.
1960s	Large scale of immunisation of children with measles, mumps and rubella vaccine reduced dramatically the number of children affected with these diseases.

Summary

◆ A **bacterial cell** consists of cytoplasm, a membrane, a cell wall and genes. There is no distinct nucleus.

◆ A **virus** is smaller than a bacterium and consists of a protein coat which surrounds a few genes.

◆ Viruses can only reproduce in living cells.

◆ Some methods by which the body defends itself from disease-producing micro-organisms include:
  – the skin acting as a barrier
  – the cells lining the respiratory tract produce mucus that traps microorganisms
  – the blood producing clots that seal cuts.

◆ Some white cells:
  – ingest microorganisms
  – produce **antibodies** that destroy particular bacteria and viruses
  – produce **antitoxins** which counteract the toxins (poisons) produced by bacteria and viruses.

◆ During immunisation, mild or dead forms of the infecting organism are introduced into the bloodstream. Some white blood cells respond by producing antibodies. These antibodies can be rapidly produced if the same strain of infecting organism invades the body on another occasion. This is immunity.

◆ Life styles are related to the spread of disease.

Topic questions

1 In what ways is a bacterium cell different from the cell of a virus?

2 Describe three ways in which the body prevents infection by bacteria or viruses.

3 Describe three ways the white blood cells defend against bacteria or viruses.

4 How does immunisation help you to be protected from further infection by the same disease?

3.6	Drugs

Co-ordinated	Modular
10.12	Mod 02
	11.7

Our senses and co-ordination can be affected if we take certain substances into the body either accidentally or deliberately. Many of these substances are **drugs**. They alter the way the body works.

Many drugs are, of course, helpful. These include pain killers such as aspirin, and antibiotics like penicillin which slow the growth of bacteria so that the body can destroy them quicker. This helps sick people to recover more rapidly.

All these drugs are useful if used correctly but can be harmful if not. Pain killers only take the pain away, they do not cure the cause of the pain. Antibiotics are becoming less effective because some bacteria are becoming resistant to them.

Some drugs are harmful because they can be misused. Some examples of these are given below.

- Alcohol is a drug because it alters the way the body works. It is a small soluble molecule which is very quickly absorbed in the mouth and stomach. This means that its effects are also very quick – it slows down the reactions of the body and leads to a loss of self control. Excess alcohol can have long-term effects including permanent damage to brain and liver cells.

- Certain **solvents** such as those used in various glues, can be very dangerous. They cause hallucinations so that a person can become violent. Like alcohol, they can cause damage to the liver and brain. The problem is that both alcohol and solvents are **addictive** which means that with prolonged or heavy use, the body feels as though it cannot do without them.

- **Nicotine** is another drug. It is taken in when the tobacco in cigarettes is smoked. It has a very powerful effect on the body and, like alcohol and solvents, it is addictive, which makes giving up smoking very difficult. Babies born to mothers who smoked during pregnancy are sometimes born with nicotine in their blood.

Did you know?

The effect of an absence of nicotine after smokers have 'quit' can be reduced by wearing 'patches' containing small amounts of the drug. The body absorbs the drug through the skin. Gradually the amount in the patches is reduced so that the person has less nicotine in the blood and this lessens the addiction to it.

The effects of smoking

The passages in the nose cavity are lined with tiny hairs called cilia and the cells produce mucus (see section 3.5). Dust and microbes get caught in the sticky mucus and the cilia. The particles are expelled from the body during sneezing.

The tobacco smoke from cigarettes can stop the cilia beating and this allows the mucus to build up. The only way the body can get rid of the mucus, the dust it has trapped and bacteria is to cough it up. The smoke also contains a large number of chemicals, including tar, and some of these are known carcinogens (**cancer**-causing chemicals). Frequent coughing damages the lungs and can lead to severe lung diseases, such as bronchitis and emphysema. Emphysema results from the breakdown of the thin membranes of the alveoli and results in a reduction of gaseous exchange surface. This damage eventually prevents the lungs from working properly and the person may become disabled by breathing difficulties.

Another problem caused by smoking is the carbon monoxide produced by the burning cigarette. Carbon monoxide joins irreversibly to haemoglobin and prevents it carrying oxygen.

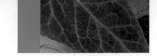

A reduction in the amount of oxygen carried by the blood is particularly dangerous to the fetus carried by a smoking mother as the baby is often born underweight. The cause of 'low birth weight' babies to smoking mothers could be that the fetus receives less oxygen for respiration.

Figure 3.29
The facts about smoking

Here are the facts for you to make up your own choice about smoking.

Nicotine:
Dose – per cigarette – 3 mg of which 1 mg is absorbed and reaches the brain in 30 seconds

Effect – increases heart rate and blood pressure putting extra strain on heart and capillaries in lungs, kidney etc.; causes stickiness of platelets leading to blood clotting which can occur anywhere in the body

Psychological effects – smoker feels more relaxed and capable of facing stressful situations; makes boring times more tolerable

Nicotine is addictive both psychologically and physiologically

Carbon monoxide:
Dose – 5% of cigarette smoke is CO

Effect – lowers oxygen-carrying capacity of blood haemoglobin by 3 – 7% permanently. This is serious for people with heart problems or for pregnant mothers as their fetus receives less O_2 and grows slowly

Tars:
Dose – enters lung as an aerosol in tobacco smoke. 70% is deposited in bronchioles and alveoli

Effect – tar causes chronic bronchitis and carcinogens greatly increase chances of developing lung cancer, cancer of the mouth and cancer of the throat

Irritants: including tar
Effect – cause extra secretion of mucus, coughing, breakdown of alveolar walls

The links between smoking and lung cancer

In 1585 Sir Francis Drake brought tobacco to England and Sir Walter Raleigh introduced the practice of pipe smoking among the Elizabethan courtiers. For more than 300 years tobacco was smoked only in pipes or as cigars. It was not until early in the 20th century did the smoking of cigarettes become common. By 1935 cigarette smoking was the most popular way of smoking tobacco and the smoking of cigarettes was socially acceptable. Up until the late 1950s the majority of the population were smokers.

Because medical records in the 19th century showed that deaths from lung cancer were very rare, no connection was made linking lung cancer to the smoking of tobacco. However as deaths from lung cancer began to increase some scientists began to look for links with the increase of cigarette smoking. Today, lung cancer kills about as many people each year as died from *all* forms of cancer in 1900.

In the 1950s scientists began to carry out detailed research to prove links between the incidence of lung cancer and smoking. The results of one very large survey are shown in the table.

	Number in sample	Number of deaths from lung cancer	Lung cancer rate per 100 000 people
Non smokers	32 460	2	6.0
Smokers	107 897	152	145

These results show that smokers increase their chances of dying from lung cancer by 24 times. Further research has shown that smokers who smoke at least 40 cigarettes a day increase their chances of dying from lung cancer by 90 times.

Because much of the research carried out *does* link an increase in deaths from lung cancer to smoking, TV tobacco commercials have been banned, tobacco products have health warnings on them, and smoking is banned in many public places. However governments in many countries collect a very large amount of money from taxes on tobacco. So rather than prohibit smoking most governments prefer to provide their people with enough information to make their own minds up about whether to smoke or not.

Summary

- Harmful **drugs** change the chemical processes in a person's body so that he or she may become dependent or addicted to them and suffer withdrawal without them.

- **Solvents** affect behaviour and may damage the lungs, liver and brain.

- Alcohol affects the nervous system by slowing down reactions which can lead to a lack of self-control, unconsciousness or even coma.

- Alcohol can damage the liver and brain.

- Nicotine is the addictive substance in tobacco.

- Tobacco contains substances which can cause lung cancer, bronchitis, emphysema, diseases of the heart and blood vessels.

- Tobacco smoke contains carbon monoxide which reduces the oxygen-carrying capacity of the blood. In pregnant women this can lead to babies being underweight.

- The link between smoking and lung cancer only gradually became accepted.

- Carbon monoxide combines irreversibly with the haemoglobin in the red blood cells.

Topic questions

1. Name two organs of the body damaged by drinking too much alcohol.

2. Name three organs of the body damaged by solvent abuse.

3. What is the name of the addictive substance in tobacco?

4. Drugs such as alcohol, solvents and nicotine can be addictive. What does this mean?

5. Why should pregnant women not smoke?

6. What effect does smoking have on the cilia that line the air passages of the respiratory system?

7. What gas in tobacco smoke lowers how much oxygen the red blood cells can take around the body?

8. Explain why alcohol acts so quickly after it has been taken into the body whereas most other drugs take about 4 hours before they act.

9. Explain why it is dangerous to drive a motor car after even a small amount of alcohol has been taken in.

10. Why is carbon monoxide such a dangerous gas to breathe in?

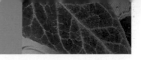

Examination questions

1 The chart shows the effect of smoking on the annual death rate in men.

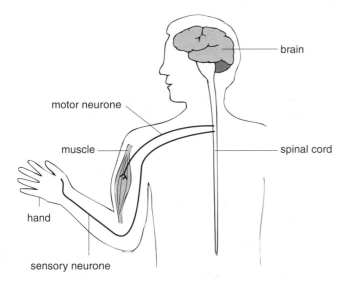

a) The death rate of men aged 45–54 who smoke more than 25 cigarettes a day is higher than the death rate for non-smokers. How much higher is it? Give your answer as a number per 100 000. *(1 mark)*

b) Explain, as fully as you can, why the death rate for smokers is higher than the death rate for non-smokers in each age group. *(3 marks)*

2 The diagram shows a reflex pathway in a human.

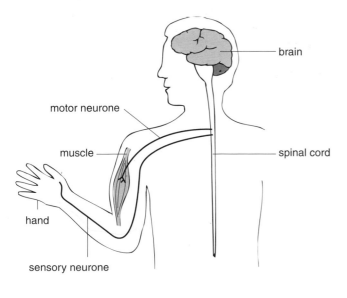

a) Label the *receptor* on the diagram. *(1 mark)*

b) Label the *effector* on the diagram. *(1 mark)*

c) i) Suggest a stimulus to the hand that could start a reflex response. *(1 mark)*

ii) Describe the response that this stimulus would cause. *(1 mark)*

d) Put arrows on the diagram to show the direction of the path taken by the nerve impulses. *(1 mark)*

3 The diagram shows the human skin.

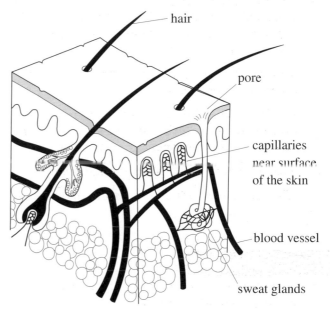

a) i) In hot weather the diameters of the blood vessels supplying the capillaries in the skin change. Explain how this change keeps us cool. *(3 marks)*

ii) Give **one** other way that the skin can keep us cool. *(1 mark)*

b) Give **one** other function of the skin. *(1 mark)*

4

> **Coordination of the body can be affected by chemicals called hormones**

a) i) Where are hormones produced? *(1 mark)*

ii) How do hormones move around the body? *(1 mark)*

b) Insulin and glucagon are hormones.

i) Where are insulin and glucagon produced? *(1 mark)*

ii) Explain the roles of insulin and glucagon in controlling blood sugar levels. *(6 marks)*

c) A hormone can be used to treat infertility in women. Explain how. *(2 marks)*

5 The diagram shows some of the processes which control the composition of blood.

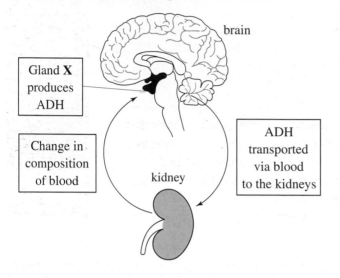

a) i)   Name gland **X** *(1 mark)*

   ii) What is the stimlus which causes gland **X** to produce ADH? *(1 mark)*

   iii) What type of substance is ADH? *(1 mark)*

b) Describe the effect of an increase in ADH production on the kidney and on the composition of the urine. *(3 marks)*

Chapter 4
Green plants as organisms

4.1		Plant nutrition

Co-ordinated	Modular
10.13	Mod 02
	11.2

One of the most important differences between green plants and animals is the way they feed. Plants can make their own food from simple raw materials. Animals have to obtain their food by eating plants and other animals. The process by which plants make their own food is called **photosynthesis**.

Photosynthesis is a process in green plants which produces **biomass** from the simple raw materials of carbon dioxide and water. The biomass is initially in the form of carbohydrate such as glucose, but from this, plants can make every substance they need. Glucose is stored in the plant as starch. The chemical reactions which convert the simple raw materials into more complex ones need a source of energy. This energy is supplied by light from the Sun which is absorbed by **chlorophyll**, a green pigment found in the chloroplasts of all plant cells which carry out photosynthesis.

Did you know?

Only 1% of the light from the Sun absorbed by the plant is used in photosynthesis.

Photosynthesis can be summarised as the word equation:

carbon dioxide + water [+ light energy] ⟶ glucose + oxygen

The need for carbon dioxide, water, light and chlorophyll in photosynthesis can be investigated by depriving plants of each of those materials in turn and seeing whether they can still make starch. Other experiments can show where the carbon dioxide goes when plants make starch and where the oxygen comes from. These experiments usually require the use of radioactive materials.

In most plants, photosynthesis takes place in the leaves. Leaves are adapted to absorb as much light as possible and to enable the plant to convert the maximum possible amount of light into chemical energy.

Carbon dioxide enters the leaves by **diffusion**. It moves from a higher concentration, outside the leaf, to a lower concentration, inside the leaf (see section 1.4).

Adaptation of the leaf for photosynthesis

- The leaves are flat and thin to provide a large surface area for the absorption of light.

- **Palisade cells**, which contain the most chloroplasts, are near the upper surface of the leaf. This enables the chlorophyll to absorb as much of the light as possible.

- Carbon dioxide enters and oxygen exits the leaf through **stomata**. **Guard cells** control the opening and closing of the stomata and therefore control the movements of gases into and out of the leaf. In most plants, stomata open when it is light, and close when it is dark.

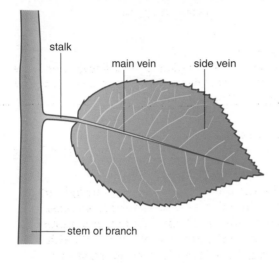

Figure 4.1
The structure of a leaf

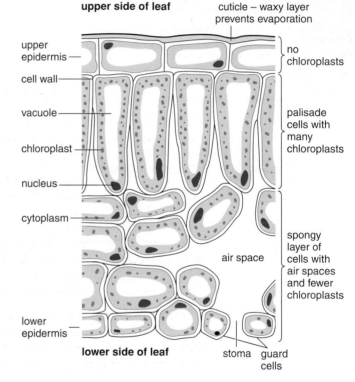

Figure 4.2
Cells in the cross section of a leaf

Limiting factors

Since photosynthesis needs the raw materials carbon dioxide, water and light energy, it is clearly going to be affected by their availability. They become **limiting factors**, i.e. the amount of photosynthesis that takes place is determined by whichever raw material is in shortest supply.

It would be expected that plants make more glucose on a warm sunny day than on a cold dull one. This can be investigated by measuring how much photosynthesis is going on. There are several ways of doing this:

- measure how much glucose or starch is being produced, although the plant has to be destroyed to make the measurements.

- measure how much oxygen is being made, this can be done easily by counting the oxygen bubbles coming off from a plant in water.

- measure how much carbon dioxide is being used up. However, a plant does not always take the carbon dioxide out of the air. Some of the carbon dioxide is produced by the plant itself by respiration.

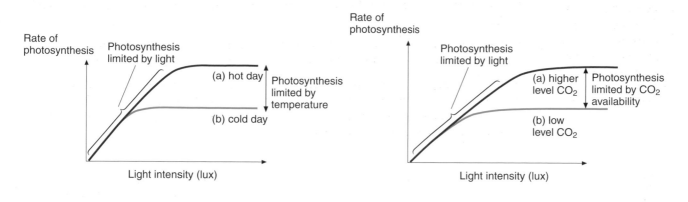

Rate of photosynthesis

Photosynthesis limited by light

(a) hot day

Photosynthesis limited by temperature

(b) cold day

Light intensity (lux)

Rate of photosynthesis

Photosynthesis limited by light

(a) higher level CO_2

Photosynthesis limited by CO_2 availability

(b) low level CO_2

Light intensity (lux)

Figure 4.3
Graph to show the effect of increasing light intensity (i.e. increasing brightness) on the rate of photosynthesis on a) a hot day and b) a cold day

Figure 4.4
Graph to show the effect of increasing light intensity on the rate of photosynthesis at a) high CO_2 and b) low CO_2 concentrations in the air

Figure 4.3 shows the effect of the increase in light intensity on the rate of photosynthesis as a day progresses on a) a warm sunny day and b) a day that is just as sunny but much colder.

The rate of photosynthesis increases with the intensity of the light. The light must have been a limiting factor. Beyond a certain point there is no further increase however bright the light becomes. This indicates that light is no longer a limiting factor but that something else is – it could be the amount of carbon dioxide. Figure 4.4 shows what happens when the effect of light is investigated at two different levels of carbon dioxide but at the same temperature. The level of carbon dioxide was the limiting factor because when the level of carbon dioxide is increased, the rate of photosynthesis again increases with light intensity.

It is possible to apply the same principle to any of the factors which affect the rate at which plants make glucose. Experiments show that water can also be a limiting factor. In dry summers and in hot climates the availability of water is likely to be the most important limiting factor of all.

Did you know?

Plants which can make their own food by photosynthesis are green because they contain chlorophyll. This is a vital substance in converting light energy into chemical energy. Chlorophyll appears green because it reflects green light. This means that it cannot absorb very much green light and must absorb mostly red and blue.

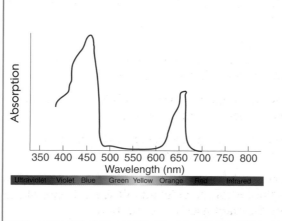

Figure 4.5
Absorption of different wavelengths of light by chlorophyll

Absorption

350 400 450 500 550 600 650 700 750 800
Wavelength (nm)

Ultraviolet Violet Blue Green Yellow Orange Red Infrared

Figure 4.5 shows how much of each colour of light is absorbed by green plants and how much photosynthesis goes on in each colour. Maximum absorption is in the blue/violet and orange/red ranges.

There is a close link or correlation between the colour that is absorbed and the colour that is used by the plant. This suggests that plants produce most food when light in the red and blue parts of the spectrum is shone at them and very little in the green part.

Uses of the products of photosynthesis

The equation for photosynthesis shows that both glucose and oxygen are made by photosynthesis. This is only part of the story, however, because plants can make everything they need from photosynthetic products. The glucose can be used to make every other substance the plant needs.

The energy released from glucose during respiration is used to build smaller molecules into larger molecules. For example:

1 Starch – Some of the glucose is made into starch and stored in the roots, stems and leaves. Starch is insoluble and can easily be stored without upsetting the water balance of cells, so the amount that can be stored by plants is almost unlimited.

2 Cellulose – Another important material made is cellulose. This is important for the manufacture of cell walls which all plant cells have. It is a very tough material and helps the plant cell walls to support the weight of the plant.

3 Lipids – Plants can convert glucose into lipids for storing in seeds. Like starch, lipids are insoluble but by weight they contain about twice as much energy as starch. A lot of energy can therefore be stored in a smaller space. The seeds need the stored energy so that they can remain dormant in the winter until conditions favourable to growth return in the spring. The lipids are then used to supply the young seedling with the energy it needs until it can make its own food.

4 Amino acids – All the substances described so far contain only the atoms carbon, hydrogen and oxygen. Plant cells can also make amino acids from glucose. Amino acids contain the element nitrogen in addition to carbon, hydrogen and oxygen, and to make them, plants must absorb nitrate from the soil. Amino acids are then made into protein, a vital building block of cells. This is why the healthy growth of plants is often dependent on the application of nitrate fertiliser to the soil.

Mineral ions

Besides needing water from the soil to photosynthesise, plants need mineral ions to make other compounds. Plants absorb these mineral ions from the soil. In farming these mineral ions are removed from the soil when the crop is harvested. To maintain crop yields, the mineral ions must be replaced by fertilisers. The most common artificial fertiliser is known as NPK because it contains nitrogen, phosphate and potassium. What does a plant need these three ions for?

Nitrate

Plants need nitrates to make proteins. Without proteins plants cannot grow. The symptoms of nitrate deficiency are stunted growth and yellowing of the older leaves.

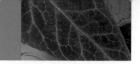

Phosphate

Plants need **phosphates** for the energy transfers that take place in photosynthesis and respiration. Without phosphate the plant cannot obtain enough energy for normal growth. The symptoms of phosphate deficiency are poor root growth and purple younger leaves.

Potassium

Potassium is needed to produce some of the enzymes involved in photosynthesis and respiration. The symptoms of potassium deficiency are yellow leaves with dead spots.

Summary

- **Photosynthesis** can be summarised by the equation:

  carbon dioxide + water + [light energy] $\longrightarrow$ glucose + oxygen

- During photosynthesis:

  - light energy is absorbed by a green substance called chlorophyll. **Chlorophyll** is found in the chloroplasts of green plants

  - the light energy is used by converting carbon dioxide and water into a sugar (glucose)

  - oxygen is released as a by-product.

- The rate of photosynthesis may be limited by such factors as:

  - low temperatures

  - a shortage of carbon dioxide

  - a shortage of light.

  These are called **limiting factors.**

- Much of the glucose produced is often converted into and stored as insoluble starch.

- Some of the glucose is used by the plant cells for respiration.

- The energy transferred during respiration is used by plants to build smaller molecules into larger molecules, such as:

  - glucose into starch

  - glucose into cellulose for cell walls

  - glucose, nitrates and other nutrients into amino acids which are then built up into proteins

    glucose into lipids (fats or oils) for storage in seeds.

- Plant roots absorb mineral salts including nitrates all of which are needed for healthy growth.

- For healthy growth plants need mineral ions such as:

  - nitrates, for the synthesis of proteins

  - phosphates, which play an important role in the energy transfers involved in photosynthesis and respiration

  - potassium, which helps to produce some of the enzymes involved in photosynthesis and respiration.

- Where these mineral ions are missing from the growing conditions the following deficiencies will be observed:

  - a lack of nitrate ions results in stunted growth and a yellowing of the older leaves

  - a lack of phosphate ions results in poor root growth and young leaves going a purple colour

  - a lack of potassium ions produces yellow leaves with dead spots.

Topic questions

1 Why are most leaves wide and flat?

2 a) Write down the word equation for photosynthesis.
   b) What are the two reactants in this equation?
   c) What are the two products in this reaction?

3 Photosynthesis takes place in cells containing chlorophyll.
   a) What colour is chlorophyll?
   b) Where in a cell is chlorophyll found?
   c) What is the function of chlorophyll?

4 The graph shows how the rate of photosynthesis changes as the light intensity changes.

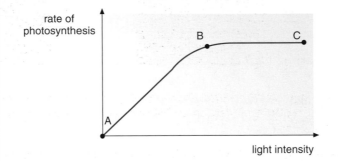

   a) What trend is shown between A and B?
   b) What trend is shown between B and C?
   c) Suggest why the rate of photosynthesis is not increasing beyond B?

5 What is some of the glucose made during photosynthesis stored as?

6 Much of the energy transferred during respiration is used by plants to build large molecules from small molecules. Name four examples of large molecules made by plants. For each example state what use it is to the plant.

7 Copy and complete the table

Mineral ion	Why needed by a plant	Effect if missing from the growing conditions
nitrate		
phosphate		
potassium		

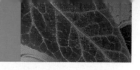

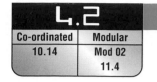

Co-ordinated	Modular
10.14	Mod 02
	11.4

Plant hormones

Plants are sensitive to light, moisture and gravity. Plant shoots grow *towards* the light and *away* from the force of gravity. The advantage of this is that the plant's leaves are more likely to be in the light where they can absorb energy for photosynthesis.

Figure 4.6
Cress plants grown with light from above (left) and with light from only one direction (right)

Plant roots grow *towards* the force of gravity and *towards* water. The advantage of this is that the roots are more likely to anchor the plant firmly in the soil and be able to absorb water.

Animals have chemicals produced by glands circulating in their bloodstream which can bring changes in cells, tissues and organs a long way from the gland which produced them. These chemicals are called **hormones**. Plants also produce hormones. One type of plant hormone is produced at the shoot tip. The hormones pass down the stem from the shoot tip and cause the cells just behind the shoot tip to elongate, making the shoot grow.

Response of plant shoots to light

In growing shoots some hormones respond to the effects of light. These are considered to encourage cell growth. If a plant is lit equally from all sides the hormones produced by the growing tip diffuse down the shoot evenly. If the plant is lit only from one side then more of the hormones produced by the growing tip diffuses down and collects on the unlit side of the growing shoot. The hormone encourages faster growth on the unlit side so as the shoot grows, it bends towards the light.

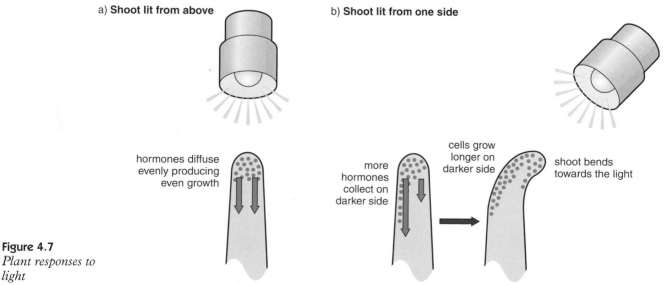

a) **Shoot lit from above**

hormones diffuse evenly producing even growth

b) **Shoot lit from one side**

more hormones collect on darker side

cells grow longer on darker side

shoot bends towards the light

Figure 4.7
Plant responses to light

Response of plant roots to gravity

In roots some of the hormones produced by the growing tip respond to the effects of gravity. These hormones slow down the growth rate of the cells in the growing tip. So if a root tip is growing downwards then the hormones are evenly distributed and the root tip continues to grow straight down. If the tip is growing sideways then more hormones collect on the underside of the growing tip, causing the cells on the underside to slow down their rate of growth. This causes the root tip to grow downwards.

Figure 4.8
Plant responses to gravity

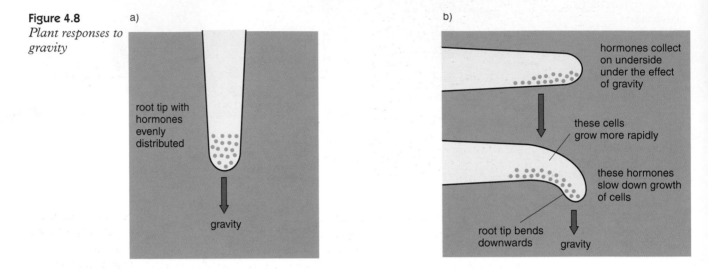

Commercial uses of plant hormones

It is now possible to make artificial plant hormones.

• If plants are given too much hormone, they can grow too quickly and die. Broad-leaved plants, such as some weeds (dandelions), are more affected by this than narrow-leaved plants, such as grasses. If the hormone is sprayed onto a lawn, the weeds are killed and the grass is left unaffected. The hormone is being used as a selective weed killer.

Figure 4.9
A dock plant before (left) and after treatment with a weed killer

Artificial plant hormones can also be used to stimulate the growth of root tissue in plants. If cut stems of plants are dipped in a powder containing small amounts of rooting hormone, the cut stem will develop roots. This is used to make sure that cuttings taken from plants develop roots more quickly and to increase the chances of the cutting developing into a full-grown plant. Cuttings are a very important way of increasing the stock of plants more quickly than sowing seeds (see section 5.4).

Figure 4.10
Cuttings dipped in rooting hormone quickly develop roots

- Hormones in plants are important in the development of fruits after fertilisation. Artificial hormones can be sprayed on the plant at flowering time so that the fruits will grow without pollination and fertilisation taking place. This leads to the production of fruits that do not contain seeds (so-called seedless fruits such as seedless grapes).

Summary

- The shoots of a plant grow *towards* the light and *against* the force of gravity.

- The roots grow *towards* water and *towards* the force of gravity.

- Growth in plants is co-ordinated by **hormones.**

- The responses of roots and shoots to light, water and gravity are due to the unequal distribution of hormones. This unequal distribution causes unequal growth rates.

- Hormones controlling the growth and reproduction in plants can be used by humans to:

  - produce large numbers of plants quickly by stimulating root growth in cuttings
  - regulate the ripening of fruit
  - kill weeds.

Topic questions

1   To what three things are plants sensitive?

2   a) Towards what do plant shoots grow?
　　b) How does this help the plant survive?

3   a) Towards what do plant roots grow?
　　b) Why is this an advantage for the plant?

4   What do plants produce that causes them to control how they grow?

5   If a plant has light coming from one side only it bends towards the light. Explain why.

6   If a root is growing sideways it will bend downwards under the effect of gravity. Explain why.

7   Give three commercial uses of plant hormones.

Transport and water relations in plants

Water is vital to plants for the following:

- Photosynthesis – The process of photosynthesis cannot proceed without water which is a raw material for the reaction.

- Transport – Plants need to transport materials, such as glucose or mineral ions, in solution from one part of the plant to another.

- Mineral ion uptake – Materials such as nitrates, phosphates and other mineral ions from the soil can only be absorbed from the soil if they are in solution.

- Support – If plants suffer from a lack of water, they may **wilt** and eventually die because the cells are no longer swollen due to a lack of **turgor** caused by a lack of water.

Transpiration

Water vapour evaporates from the surfaces of the leaves. This process is called **transpiration**. Transpiration will occur most rapidly in hot, dry and windy conditions. In order to reduce the amount of water loss many plants have a waxy surface. Plants, such as cacti, that grow in very hot, dry areas often have leaves with a very thick waxy surface and a very small surface area. The water evaporates through the **stomata**, the opening through which gases enter and exit the leaf (see section 4.1). The size of the opening is controlled by the pairs of **guard cells** that surround each stoma.

Figure 4.11
Stomata and hairs on the surface of a leaf

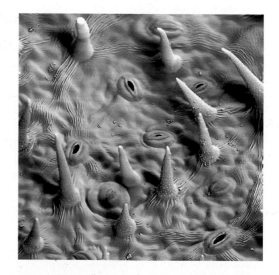

If plants lose water faster than the roots can take it from the soil the stomata will close to stop the plant from wilting.

How does water move from the roots to the leaves?

The evaporation of water from the leaves causes a transpiration pull. This is powerful enough to suck the water from the roots all the way up the plant. Water travels up the stem and into the leaves through **xylem** vessels.

Did you know?

Transpiration can raise water from the roots to the leaves of the tallest plants which can be over 100 m tall?

Did you know?

How stomata open and close

The wall of each guard cell is thickest and therefore less likely to change its shape where it lines the gap. When the vacuoles of a pair of guard cells fill with water, by osmosis, the guard cells swell and bend. They bend because the guard cells are joined at each end. This causes the gap between them to widen. When guard cells lose water they shrink and straighten. This causes the gap between them to get narrower.

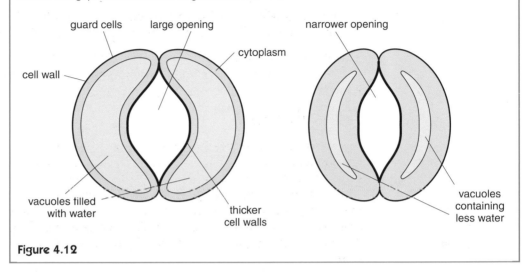

Figure 4.12

Uptake of water and mineral salts by plants

Plants take in nearly all the water and mineral salts they need through the roots. The water is absorbed by **osmosis** (see section 1.4).

The mineral salts are absorbed in solution in the form of ions e.g. NO_3^-, PO_4^{2-}, K^+ (see section 1.4).

The roots of the plants are covered with cells called **root hair cells**.

Figure 4.13
Diagram to show the role of the root hair cell in the uptake of water and the movement of water through the root to the xylem

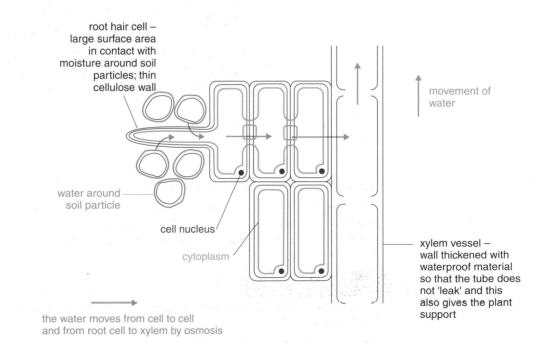

the water moves from cell to cell and from root cell to xylem by osmosis

These root hair cells have very thin walls and greatly increase the surface area for absorbing mineral salts and water from the soil.

The salts get into the root hairs by **diffusion** or by active transport (see Chapter 1). To be absorbed by diffusion, there must be a higher concentration of mineral salts in the solution outside the cell than inside. This enables a concentration gradient to exist and then mineral salts diffuse into the root hair cells down the concentration gradient.

Active transport is needed if there is a higher concentration of salts inside the cells than outside. In such cases, the cell can only take in the salts if it uses cellular energy.

Once inside the root hair cells, the water and mineral salts are passed to the **xylem** tissue in the centre of the root. Xylem contains differentiated cells that form vessels. The xylem vessels transport the mineral ions up the root and stem to all cells.

The carbohydrate made by photosynthesis must also be moved to other parts of the plant. The soluble food is transported in the **phloem** tubes to the roots for storage or to the tips of the shoots where it is used as a source of energy for growth.

Figure 4.14

The passage of water through a plant

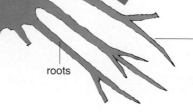
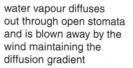

leaf

transpiration is the loss of water vapour from the leaf. As the water evaporates from the leaf, turgor in the cells drops and the sap becomes more concentrated so water is drawn from the xylem vessels by osmosis. Transpiration 'pulls' water up the stem

water vapour diffuses out through open stomata and is blown away by the wind maintaining the diffusion gradient

stem

water molecules attract one another and are also attracted to the walls of the xylem vessels. This means the water molecules are not easily pulled away from each other (they behave like a string of paper clips). The water loss from the leaf by transpiration results in the transpiration pull and causes the water to rise up the stem in the transpiration stream

xylem vessels
Sap is more concentrated than soil water solution so water passes in by osmosis. Capillary rise takes place because the vessels have a small diameter

uptake of water and mineral salts from the soil. Mineral ions are taken in by active transport by the root hairs. These mineral salts are moved into the xylem, increasing its ion concentration. Water passes from the dilute soil water through the membranes of the root hair cells to the higher concentration in the xylem vessels by osmosis.

soil water has a dilute solution of mineral salts

roots

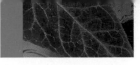

Water and the cell

Figure 4.15
a) A turgid plant cell,
b) a flaccid plant cell

a) **Turgid cell**

cell membrane
pressing against
cellulose cell wall

large vacuole filled
with cell sap

b) **Flaccid cell**

cell membrane
pulling away from
cell wall at the
corners; there is
less pressure on
the cell wall

vacuole beginning
to shrink

Cells normally contain as much water as possible. In this state the cells are said to be turgid. The vacuole in the cell which contains the sap exerts a pressure on the walls of the cell, helping to support it. When all cells are turgid this helps to support the whole plant. This is particularly important in non-woody plants.

Sometimes water evaporates from the surface of the leaf faster than it can be replaced by osmosis. This means that cells lose water faster than they can gain it. The vacuole of the plant cell begins to shrink and the pressure on the wall of the cell decreases. The cell becomes flaccid. If this happens in many cells, the leaf may begin to wilt. If the plant is given water, the cell becomes turgid again and the leaf recovers.

Summary

- **Transpiration** is the loss of water vapour from the surface of a leaf.

- Transpiration is most rapid in hot, dry and windy conditions.

- Most leaves have a waxy coating (cuticle) that prevents the loss of too much water.

- Transpiration takes place through the **stomata**.

- The size of the stomata is controlled by **guard cells**.

- Water in a plant provides support. Plants **wilt** if they lose too much water.

- **Xylem** tissue transports water and minerals from the roots to the stem and leaves.

- **Phloem** tissue transports nutrients, such as glucose, from the leaves to the rest of the plant

- Water moving into a cell by osmosis increases the pressure inside the cell.

- Cell walls are strong enough to withstand this pressure.

- This pressure keeps the cell rigid (maintains its turgor). This **turgor** provides the support.

Topic questions

1 a) What is transpiration?
 b) In which conditions would transpiration be slowest?

2 Through which part of a leaf does the water evaporate?

3 What controls how much water evaporates?

4 Why do plants wilt on a hot day?

5 What draws water all the way up from the roots to the leaves?

6 What is transported along the xylem tissues?

7 What is transported along the phloem tissues?

8 Describe how cells become rigid but do not burst.

Examination questions

1 a) Copy and complete the following sentences.

Green plants produce their own food by a process called photosynthesis. In this process the raw materials are _____ and carbon dioxide. Glucose and _____ are produced. _____ energy is absorbed by the green substance called _____ . *(4 marks)*

 b) Name **two** things that can happen in the plant to the glucose produced in photosynthesis.
  (1 mark)

 c) Plants need mineral salts.
  i) Through which part do mineral salts get into the plant? *(1 mark)*
  ii) Explain why water is important in this process. *(2 marks)*

 d) Some students set up water cultures to find out how plants use nitrates.
 They had two sets of nutrient solutions.
 A full solution provided the plant with all the required nutrients.
 The results table shows the average mass of the seedlings after 28 days growth.
  i) Give a conclusion you could make from these results. *(1 mark)*
  ii) Calculate the difference in average mass caused by the addition of nitrates to the culture solution. *(1 mark)*
  iii) What are nitrates used for in the seedling? *(1 mark)*
  iv) Some factors need to be controlled to keep this test fair. Name **two** of them. *(2 marks)*
  v) Suggest **one** way you could improve the experiment. *(1 mark)*

Culture solution	Average mass of seedling in g
distilled water	0.14
full solution with no nitrates	0.29
full solution	0.43

2 The diagram shows a plant leaf during photosynthesis.

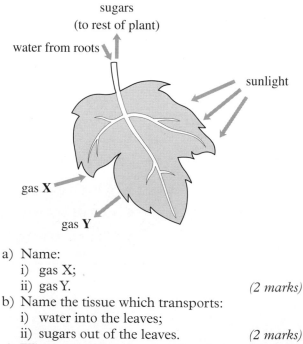

 a) Name:
  i) gas X;
  ii) gas Y. *(2 marks)*

 b) Name the tissue which transports:
  i) water into the leaves;
  ii) sugars out of the leaves. *(2 marks)*

 c) Why is sunlight necessary for photosynthesis?
  (1 mark)

d) Some of the sugars produced by
photosynthesis are stored as starch in the roots.
Explain, as fully as you can, why it is an
advantage to the plant to store carbohydrate as
starch rather than as sugar. *(3 marks)*

3 A potted plant was left in a hot, brightly lit room
for ten hours. The plant was not watered during
this period. The drawings show how the mean
width of stomata changed over the ten hour
period.

a) Why do plants need stomata? *(1 mark)*
b) Name the cells labelled **X** on the drawing.
(1 mark)
c) The width of the stomata changed over the ten
hour period. Explain the advantage to the plant
of this change. *(2 marks)*

d) Explain, in terms of osmosis, how cells in a
young plant are kept rigid. *(2 marks)*

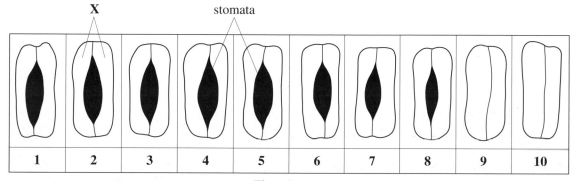

Time (hours)

Chapter 5

Variation, inheritance and evolution

Key terms adaptation • allele • artificial selection • asexual reproduction • cancer • carrier • cell division • characteristic • chromosomes • clone • DNA • dominant • extinct • fertile • fossils • fusion • gamete • gene • genetic • genetic engineering • genotype • growth • heterozygous • homozygous • meiosis • mitosis • mutation • natural selection • recessive • reproduction • selective breeding • sexual reproduction • species • variation

5.1		Variation
Co-ordinated	**Modular**	
10.16	Mod 04 13.1	

People are all different because each person has a unique set of genetic information. The only exceptions are identical twins which are formed from a single fertilised egg cell.

The children in the photograph can easily be divided into groups of males and females. But it's not so easy to group them according to variations in height, eye colour or whether they have ear lobes.

What causes variation?

Differences in the **characteristics** of individuals of the same kind (species) are caused by differences in:

- **environmental factors**. For example, plants grown in plenty of light, fertile soil and which are well watered, will grow sturdy and strong. But if plants are grown in poor light and are not watered properly they will be pale and spindly. For both plants and animals, nutrients will determine whether or not the organism reaches its genetic potential size. However, environment cannot change some inherited characteristics. Living in the dark or only eating chips will not change a person's blood group.

- **genes** they have inherited (**genetic factors**). Cell division to produce **gametes** (**meiosis**) is a source of variation as each individual receives a unique genetic combination.

- **Mutation** – A mutation is a change in the genetic information. Mutations can be of two types:

  1. a chromosomal mutation is a major change such as a piece of **chromosome** breaking off and either getting lost or attached to another chromosome, or the loss of a whole chromosome, changing the chromosome number. These changes produce serious results in animals but are less catastrophic in plants.

  2. a gene mutation in which the information within a gene is changed.

Figure 5.1
Variation as seen in a group of children

A mutation that happens when the gamete is formed will be passed on to the offspring. Queen Victoria received a mutation that she passed onto some of her sons resulting in them having haemophilia. In this condition, the blood is very slow to clot.

What causes mutations?

Mutations can happen spontaneously, but exposure to gamma radiation, strong X-rays, ultraviolet rays and certain chemicals can trigger mutations. Food additives and household chemicals are always tested to try to make sure that they do not cause mutations. Excess exposure to sunlight can cause mutations in skin cell nuclei and cause skin **cancer**. A cancer develops when cells multiply in an uncontrolled way.

Are mutations harmful?

Most mutations are harmful if they occur in reproductive cells. This is because the young may develop abnormally or even die at an early stage of their development. Mutations are also harmful if they occur in body cells because they are likely to develop into cancers.

Some mutations do not affect the individual. In some cases mutations may increase the chances of an individual surviving. If such a mutation is genetic and therefore heritable, then offspring produced by such an individual are likely to have their chances of survival increased.

Variation and reproduction

There are two forms of reproduction:

- **sexual reproduction**; which involves the **fusion** of male and female gametes

- **asexual reproduction**; where there is no fusion of cells and only one individual gives rise to offspring.

Asexual reproduction produces offspring with genetic information that is identical, not only to each offspring, but is also identical to that of the parent. **Genetically** identical individuals are called **clones**.

Sexual reproduction produces offspring that have a mixture of genetic information from both parents. These individuals show variation between themselves and each of the parents.

Differences caused by the environment

The potential to grow to a particular size is determined by the genetic material present at fertilisation or, in the case of an asexually produced organism, by the genetic material present in the parent. However whether an individual reaches their potential is due mainly to the effects of environmental factors such as diet, living conditions etc. Plants show the importance of the environment very clearly. Seeds sown in poor soil will certainly not grow and develop their full potential.

Variation in hydrangeas

> **Did you know?**
>
> There is a plant called a hydrangea that has flowers that can be blue, white or pink. The colour of the flower is not controlled by genes but by the type of soil in which the plant grows. If the soil is acidic then the flowers are blue, if the soil is alkaline then the flowers are white or pink.

Why does reproduction sometimes produce variation?

The cells of offspring produced by asexual reproduction are produced by mitosis from the parental cells (see section 1.5). Because mitosis produces identical genes the offspring contain the same genes as the parent.

Sexual reproduction gives rise to variation because:

- there is a random assortment of the chromosomes during meiosis (see section 1.5) so the gametes do not contain identical genetic material,

- there is random fertilisation – it is pure chance which gametes fuse with each other, therefore the offspring have a new assortment of genetic material.

The work of Gregor Johann Mendel (1822–1884)

Mendel was an Austrian monk who was very interested in plant breeding. He was particularly interested in breeding pea plants which showed clear variation in a number of characteristics. For example, some pea plants had short stems and others had long stems, some plants produced white flowers others produced red flowers, some plants had round seed pods others had wrinkled seed pods. He started his investigations in 1856 and grew the pea plants in the garden of the monastery.

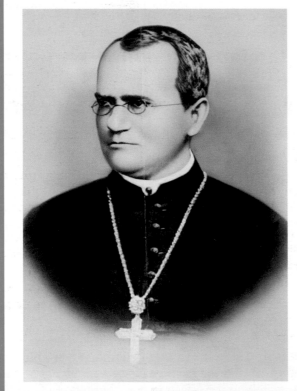

Figure 5.2
Gregor Mendel

He cross-pollinated those plants that showed the characteristics in which he was interested, taking the greatest care to ensure that only he and not the insects or the wind pollinated the flowers. Each year he collected the seeds from his selected plants and grew them. When the new plants grew he recorded their characteristics and by 1863 he had bred over 25 000 pea plants. For seven years he had taken very careful notes of all the crosses he had made together with the results of those crosses.

In the 19th century the accepted view was that the characteristics of offspring are the result of a blending of the characteristics of the parents. Mendel gradually realised that the idea of characteristics being inherited through a blending process did not match his results. He proposed the following two ideas:

- That there were hereditary factors (his word for genes) that showed either dominant or recessive properties. When a pollen grain meets an egg, a factor pair is formed in which a dominant factor will mask a recessive factor.

- That hereditary factors do not blend but remain unchanged from one generation to the next.

These two ideas form the basis of what we now know as the laws of heredity. In 1866 Mendel published his results and conclusions in a local scientific journal. For the next 34 years very few scientists took any notice of his work because:

- the journal in which it appeared was not read by many scientists

- the complexity of his results meant that the work was difficult to understand.

Mendel died with his contribution to genetics unrecognised. It was not until 1900, when a Dutch scientist came across the journal in which Mendel's work was published, that the great importance of Mendel's conclusions was realised. Several years later Mendel's ideas were used to help explain Darwin's theory of evolution.

Summary

♦ All young resemble their parents because of genetic information passed on through the sex cells (**gametes**).

♦ The information is carried by **genes**.

♦ Different genes control different characteristics.

♦ Mendel proposed the idea of genes.

♦ Differences in characteristics within a species may be due to differences in:

- the genes inherited (genetic causes)
- the conditions in which they have developed (environmental causes)
  or a combination of both.

♦ New forms of genes (**mutations**) result from changes in existing genes.

♦ Mutations can occur naturally, but can occur more frequently by exposure to:

- ionising radiation and from the radiation from radioactive substances
- certain chemicals.

♦ Most mutations are harmful if they occur in:

- reproductive cells
- body cells.

♦ Some mutations are neutral in their effects.

♦ In rare cases, a mutation may increase the chances of survival of an organism and any offspring if the mutation is genetic.

♦ There are two forms of reproduction:

- **sexual reproduction**; which involves the **fusion** of male and female gametes
- **asexual reproduction**; where there is no fusion of cells and where only one individual gives rise to offspring.

♦ Asexual reproduction produces offspring with genetic information that is identical to that of the parent. **Genetically** identical individuals are called **clones**.

♦ The cells of offspring produced by asexual reproduction are produced by mitosis from the parental cells. They contain the same genes as the parent.

♦ Sexual reproduction produces individuals that have a mixture of genetic information from both parents. These individuals show variation between themselves and each of the parents.

♦ Sexual reproduction gives rise to variation because:

- the gametes are produced from parental cells by meiosis
- when gametes fuse, one of each pair of alleles comes from each parent
- the alleles in a pair may vary and therefore produce different characteristics.

♦ Mendel's discovery about separately inherited factors only gradually came to be accepted.

Topic questions

1 What part of a nucleus carries the genetic information for a particular characteristic?

2 a) Why are people different?
   b) Which people will have identical genes?
   c) What does a single gene control?

3 a) Which type of reproduction produces identical offspring?
   b) What is the name given to the offspring that are genetically identical to each other and to the parent?

4 a) What is a gene mutation?
   b) Give four different ways a mutation can occur.

5 Give two reasons why sexual reproduction produces offspring with different genetic material?

6 Why is it harmful if a mutation occurs in
   a) the reproductive cells
   b) body cells?

7 When could a mutation be useful?

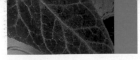

Genes

Sex determination

Humans have 23 pairs of **chromosomes** in body cells which means that the egg and sperm cells each contain 23 single chromosomes.

Of the 23 pairs, one pair is made up of the sex chromosomes that determine the sex of the person. These are labelled X and Y on Figure 5.3. Males have an X and a Y chromosome (as in the diagram) while females have two X chromosomes, XX.

Figure 5.3

A set of male human chromosomes

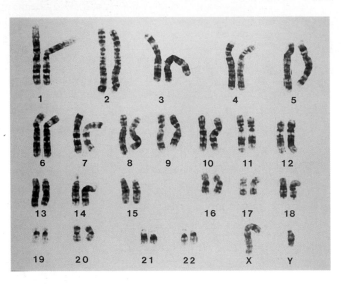

Will the baby be a boy or a girl?

- A sperm will contain either one X sex chromosome or one Y chromosome.

- Every egg will contain one X sex chromosome.

During fertilisation one sperm will fuse with one egg. After fertilisation the cells of the baby will contain two sex chromosomes. So the possible combinations for the baby's sex chromosomes will be:

	sperm		egg	
sex chromosomes	X or Y		X or X	

baby's sex chromosomes	X + X	or	Y + X
	(from a sperm) (from an egg)		(from a sperm) (from an egg)
	a baby girl		**a baby boy**

So there is a 50% chance of a baby being a boy or a girl.

Figure 5.4
Sex determination in humans

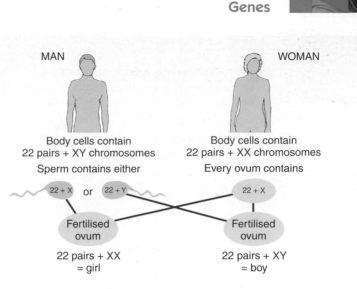

From now on only the X and Y chromosomes will be considered, although it must be remembered that there are 22 other chromosomes in each ovum and sperm.

The only possible combinations are XX and XY as shown below.

Figure 5.5
A Punnett square crossing XX and XY

		Mother's gametes		
		X	X	
				Offspring
Father's gametes	X	XX	XX	50% girls
	Y	XY	XY	50% boys

As a result it can be seen that the chances of having a baby boy or a baby girl are equal. Again there is an equal chance at each conception. However, not all families will have equal numbers of sons and daughters.

Genes, chromosomes and alleles

Chromosomes are made of long molecules of a substance called **DNA**. Each chromosome is divided into genes. A gene is a length of chromosomal DNA which codes for one characteristic. It is helpful to use the word gene to refer to the position of the information on the chromosome.

Alleles can be considered to be different forms of a gene found in the same position on the chromosome. They control the same characteristic but code for a different detail. There are very many alleles in plants and animals, each controlling different features of the organism. It is much easier to understand with an example.

Alleles for eye colour

A chromosome has a gene position for eye colour but the allele at that position could be either the allele coding for blue eyes or the allele coding for brown eyes.

Figure 5.6a

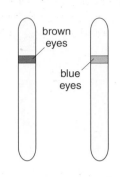

This individual has inherited the allele for brown eyes from one parent and that for blue eyes from the other parent.

- The alleles for eye colour are different so the individual is said to be heterozygous for eye colour.

- The allele for brown eyes is **dominant** over the allele for blue eyes so the instruction 'have brown eyes' hides, but does not destroy the instruction 'have blue eyes'. The individual will therefore have brown eyes.

Figure 5.6b

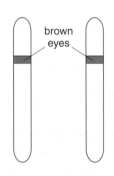

This individual has inherited one of the alleles for brown eyes from each parent.

- The alleles for eye colour are the same so the individual is said to be homozygous for eye colour.

- Because both alleles are for the same instruction the individual will have brown eyes. The individual would have blue eyes if both alleles had been for this instruction. An allele which controls the development of characteristics only if the dominant allele is not present, is called a **recessive** allele. The allele for blue eye colour is recessive.

Some dominant and recessive characteristics in humans

Characteristic	Dominant	Recessive
Eye colour	brown eyes	blue eyes
Freckled skin	freckles	no freckles
Tongue-rolling	can tongue roll	cannot tongue roll

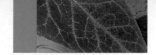

Using genetic diagrams to find out more about inheriting eye colour

- The word genotype describes the genetic composition of the organism – that is the alleles present for the characteristic being studied. When setting out a cross, the first letter of the phenotype is usually used as a capital letter to represent the dominant allele and as a lower case letter to represent the recessive allele.

Figure 5.7
A Punnett square showing the results of a homozygous cross between brown eyes (BB) and blue eyes (bb)

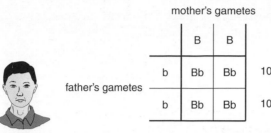

But what if the brown-eyed parent is heterozygous. This cross – heterozygous with homozygous recessive – gives a 1:1 ratio in the offspring.

Figure 5.8
A Punnett square showing the results of a cross between brown eyes (Bb) and blue eyes (bb)

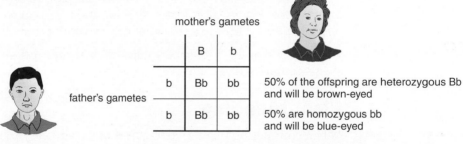

If the person has blue eyes, their genotype *must* be bb, as the allele for blue eyes is homozygous recessive. Consider what the outcome might be if a brown-eyed man and his brown-eyed wife have children. They do not know if one or both of them are heterozygous for eye colour (Bb) or homozygous (BB).

Figure 5.9
A Punnett square showing the results of a homozygous cross between brown eyes (BB)

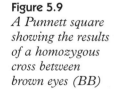
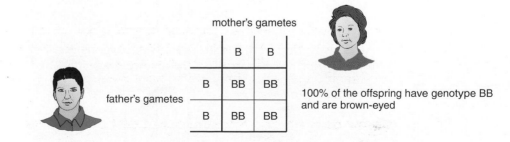

If one parent is homozygous brown eyed and the other is heterozygous, they will both look the same and the offspring will all look the same but look at the ratio of the offspring genotypes.

Figure 5.10
A Punnett square showing the results of a cross between brown eyes (BB) and brown eyes (Bb)

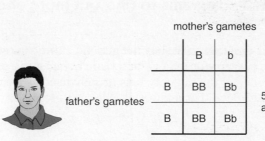

mother's gametes

	B	b
B	BB	Bb
B	BB	Bb

father's gametes

50% are homozygous brown-eyed (BB) and 50% heterozygous brown-eyed (Bb)

The third possibility is that both parents are heterozygous, and this gives the important 3:1 ratio in the results.

Figure 5.11
A Punnett square showing the results of a heterozygous cross between brown eyes (Bb) and brown eyes (Bb)

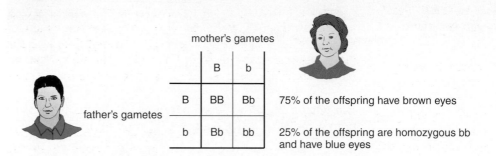

mother's gametes

	B	b
B	BB	Bb
b	Bb	bb

father's gametes

75% of the offspring have brown eyes

25% of the offspring are homozygous bb and have blue eyes

This means that there is a 75% (3 in 4) chance of these two heterozygous brown-eyed parents having a brown-eyed child, and a 25% (1 in 4) chance of offspring with blue eyes. This gives the ratio 3:1 brown:blue.

Remember that the % chance applies at each conception, so it is possible for all of their children to have blue eyes.

Plant breeding

When considering plant breeding, it is quick and easy to see the results of crosses (when one plant fertilises another). When the parents are pure homozygous RR (red) and rr (white) the following result is obtained.

Figure 5.12
A Punnet square showing the results of a cross between a red plant, RR, and a white plant, rr

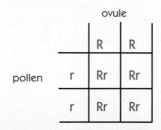

ovule

	R	R
r	Rr	Rr
r	Rr	Rr

pollen

All the offspring from this cross are heterozygous, Rr, and will have red flowers. If the offspring are then crossed, the following result is obtained.

Figure 5.13
A Punnet square showing the results of a cross between Rr and Rr

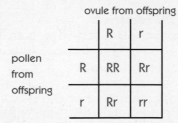

ovule from offspring

	R	r
R	RR	Rr
r	Rr	rr

pollen from offspring

In this generation, 50% of the offspring are heterozygous, Rr, 25% are homozygous RR and 25% are homozygous, rr. Therefore 75% are red flowers and 25% are white flowers, a 3:1 ratio.

Genetic diseases

Cystic fibrosis

Cystic fibrosis is an inherited disorder of the lungs and digestive system. In cystic fibrosis the mucus made in the lungs is abnormally thick which results in more lung infections, coughing and wheezing. The problems with digestion occur because the duct that transports the digestive enzymes from the pancreas to the digestive system becomes blocked by sticky mucus. Because of this, a person with cystic fibrosis must take supplementary enzymes. The varied conditions that are recognised as cystic fibrosis are all the result of a recessive mutation in one allele.

Inheriting cystic fibrosis

The allele carrying the instruction 'have cystic fibrosis' can be called '**c**'. This allele is recessive. The allele for the instruction 'do not have cystic fibrosis' can be called '**C**'. This allele is dominant.

A person will only inherit one instruction from each parent. The possible combinations in an individual are **CC**, **Cc**, or **cc**.
An individual with the combination **CC** will **not** have cystic fibrosis.
An individual with the combination **Cc** will **not** have cystic fibrosis. However this individual is described as a **carrier**, because they still contain the instruction 'have cystic fibrosis' which he or she could pass onto the next generation.
An individual with the combination **cc** will have cystic fibrosis because the dominant allele is not present.

Figure 5.14
A Punnett square showing the results of a cross between two heterozygous individuals who are carriers of the c allele

Let C = normal allele
and c = allele for cystic fibrosis

		Mother	
		C	c
Father	C	CC	Cc
	c	Cc	cc

Result: 50% chance offspring will be heterozygous carriers of the c allele
25% chance offspring will be homozygous normal
25% chance offspring will be homozygous for the c allele and will have the disease

Therefore, if both parents are carriers, there is a 1 in 4 chance at every pregnancy that they will have a child with cystic fibrosis.

Sickle cell anaemia

Sickle cell anaemia is an inherited disorder of red blood cells. In a healthy person, haemoglobin in red blood cells binds with oxygen from the air we breathe into our lungs and carries it to all parts of the body. These cells are able to squeeze through tiny blood vessels because they are soft and round.

People with sickle cell anaemia make an abnormal form of haemoglobin which causes the red blood cells to be sickle shaped. These sickle-shaped red blood cells become stiff and distorted so they can block small blood vessels. When this occurs, less oxygen-carrying blood can reach that part of the body, causing considerable pain and tissue damage.

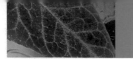

The allele that causes sickle cell anaemia is the result of a mutation. The allele is described as being abnormal. For a child to develop the disease it must have inherited the abnormal allele from both parents – making the child homozygous for the abnormal gene.

Children who are heterozygous have what is called the sickle cell trait. Their blood will contain a mixture of the two types of haemoglobin. In countries where malaria is common, this mixture provides some protection because the malarial parasite cannot survive in red cells containing abnormal haemoglobin. About 1% of the heterozygous individual's red cells are sickled; this compares with 5% of red blood cells in a person who is homozygous for the abnormal gene.

Inheriting sickle cell anaemia

The allele carrying the instruction 'do not have sickle cell anaemia' can be called '**Hb^A**'. This allele is dominant. The allele for the instruction 'have sickle cell anaemia' can be called '**Hb^S**'. This allele is recessive.

A person will only inherit one instruction from each parent. The possible combinations in an individual are **Hb^A** and **Hb^A**, **Hb^A** and **Hb^S**, or **Hb^S** and **Hb^S**. An individual with the combination **Hb^A** and **Hb^A** will **not** have sickle cell anaemia. An individual with the combination **Hb^A** and **Hb^S** will **not** have sickle cell anaemia, because the instruction 'do not have sickle cell anaemia' is dominant. However, this individual will be a carrier and will have the sickle cell trait. An individual with the combination **Hb^S** and **Hb^S** has two copies of the 'sickle cell anaemia' gene and will develop the disease.

For the Punnett square, let **Hb^A** represent the allele for normal haemoglobin, and let **Hb^S** represent the allele for abnormal haemoglobin.

Mother's gametes

		Hb^A	Hb^S
Father's gametes	Hb^A	Hb^{A}Hb^A	Hb^{A}Hb^S
	Hb^S	Hb^{A}Hb^S	Hb^{S}Hb^S

Result: 25% (1 in 4) Hb^{A}Hb^A normal haemoglobin
50% (2 in 4) Hb^{A}Hb^S sickle cell trait – these will have mostly normal haemoglobin
25% (1 in 4) Hb^{S}Hb^S these will have full sickle cell anaemia

Huntington's disease

Huntington's disease is a disorder of the nervous system. It is caused by a dominant allele of a gene and can therefore be passed on by only one parent who has the disorder.

Inheriting Huntington's disease

The allele carrying the instruction 'have Huntington's disease' – call this allele '**H**', is dominant. The allele for the instruction 'do not have Huntington's disease' – call this '**h**', is recessive.

A person will only inherit one instruction from each parent. The possible combinations in an individual are **HH**, **Hh**, or **hh**. An individual with the combination **HH** will have Huntington's disease. An individual with the combination **Hh** will have Huntington's disease, because the instruction 'have Huntington's disease' is dominant. An individual with the combination **hh** will **not** have Huntington's disease.

Figure 5.15
The red cells at the top of the photo have a sickle shape

Did you know?

Normal haemoglobin red blood cells live about 120 days before being replaced.

Figure 5.16
A Punnett square showing the results of a cross between two people with the sickle cell trait, Hb^{A}Hb^S

Let **H** = allele for Huntington's disease and **h** = normal allele.

Figure 5.17
A Punnett square showing the results of a cross between one individual who is heterozygous for the Huntington's disease gene and one individual who does not carry the gene at all

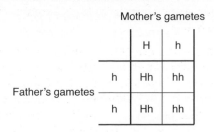

Therefore if only one parent is a sufferer, there is a 50% chance of a child being **Hh**.

Summary

◆ In human body cells one of the 23 pairs of chromosomes carries the genes that determine sex.

◆ In females the sex chromosomes are the same (XX).

◆ In males the sex chromosomes are not the same (XY).

◆ Most characteristics are controlled by one gene.

◆ Some genes have different forms called **alleles**.

◆ An allele which controls the development of a characteristic when it is present on only one of the chromosomes is the **dominant** allele.

◆ An allele which controls the development of a characteristic only if the dominant allele is **not** present is a **recessive** allele.

◆ If both chromosomes in a pair contain the same allele of a gene, the individual is **homozygous** for that allele.

◆ If both chromosomes in a pair contain the different alleles of a gene, the individual is **heterozygous** for that allele.

◆ Huntington's disease (a disorder of the nervous system) is caused by a dominant allele.

◆ A recessive gene causes cystic fibrosis (a disorder of cell membranes). Parents can be carriers without actually having the disorder.

◆ Sickle cell anaemia is a disorder of the red blood cells which reduces their oxygen-carrying capacity. Carriers of this disorder are less likely to be affected by malaria.

◆ Genetic diagrams can be constructed and used to predict and explain the outcomes of crosses for each possible combination of dominant and recessive alleles of the same gene.

Topic questions

1 a) Which pair of sex chromosomes are found in human males?
   b) Which pair of sex chromosomes are found in human females?
   c) Which sex chromosomes could be in a sperm?
   d) Which sex chromosome is in all eggs?

2 The chances of a baby being a boy or a girl are 50%. Explain why.

3 The allele for 'have freckles' is dominant over the allele for 'do not have freckles'. What does this mean?

4 a) Name a disease caused by a dominant gene.
   b) Name two diseases caused by a recessive gene.

5 A person could be a carrier of a disease. What does this mean?

6 How is being a carrier of sickle cell anaemia beneficial?

7 In a pea plant the allele for being a tall plant (T) is dominant over the allele for being a dwarf plant (t).

   a) Explain using a genetic diagram what plants would be produced if:

      i) a homozygous tall plant is crossed with a homozygous dwarf plant.
      ii) two plants from the crossing in (i) were crossed.

   b) Why is it impossible to have a heterozygous dwarf plant?

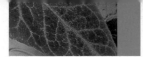

5.3 DNA

Co-ordinated	Modular
10.17	Mod 04
	13.4

Chromosomes carry a sequence of genes (Figure 5.18). A **gene** is a length or section of the chromosomal **DNA**. It codes for one **characteristic**.

The chromosomes are made up of two chains of DNA wrapped around a protein molecule and look like a twisted ladder (Figure 5.19). DNA can be considered as a polymer of base monomers (see section 10.4).

Each group of three bases is a code for a particular amino acid. So the sequence of bases in DNA controls the sequence of amino acids in a protein and therefore the type of protein produced.

The 'code' in each gene is made by the order of the chemical molecules called bases. These are represented by the coloured bars and their initial letters A, C, T, G.

Each 'gene code' has instructions for making proteins such as enzymes, hormones, pigment or protein structures. When this code is translated, we see the results as a physical characteristic such as eye colour or the production of a hormone such as insulin.

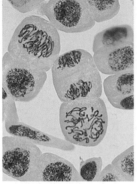

Figure 5.20
These chromosomes are visible in a slide of the tip of a root using a school microscope

Did you know?

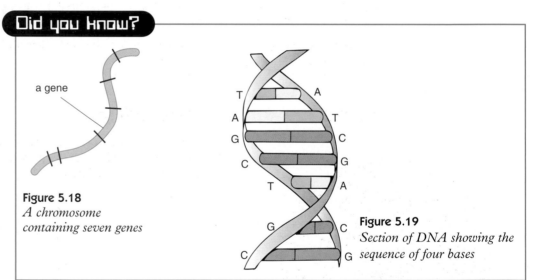

a gene

Figure 5.18
A chromosome containing seven genes

Figure 5.19
Section of DNA showing the sequence of four bases

Summary

◆ Chromosomes have long molecules of a substance called DNA.

◆ A gene is a section of DNA.

◆ DNA contains coded information that determines inherited characteristics.

◆ DNA is made of long strands made up from four different compounds called bases.

◆ A sequence of three bases is the code for a particular amino acid.

◆ The order of the bases controls the order in which amino acids are assembled to make a particular protein.

Controlling inheritance

Reproduction is the process of making new individuals. Asexual reproduction is the term used when a single individual produces offspring that are genetically identical to each other and the parent.

Cloning plants from cuttings

Florists or garden centres often display rows of almost identical plants. These were probably produced by asexual reproduction from one parent and therefore the parent plant and the new plants produced from it are genetically identical. If cuttings are taken from a favourite plant, such as a Fuchsia or African Violet, and rooted, the cuttings should grow and have exactly the same leaf shape and flower colour as the parent plant.

Such plants are called **clones** and are genetically the same as the parent plant.

Did you know?

Human cells can be cloned in tissue culture and these can be used for testing new medical drugs. Clones of human cells are a substitute for animal testing in early stages of the development of the new medicine.

Clones of skin cells can be used to grow new skin for burn victims and research is being done to grow clones of cartilage for joint repairs

Cloning plants – tissue culture

The cloning of plants through tissue culture involves the taking of a very small piece of plant and placing it in a jelly containing various nutrients.

Within a short time a layer of thick new tissue forms.

Cells are removed from this new tissue and placed in a different jelly, this time it contains not only nutrients but also hormones that speed up the growth of roots and shoots.

Once roots and shoots have developed the tiny plant is moved to a greenhouse where further growth continues.

Figure 5.21
A dish of plant cells can be the equivalent of a field of thousands of plants

Selective breeding (or artificial selection) in plants

The aim of plant breeding is to produce plants with particular desirable characteristics selected from existing plants of the same species. For example, a specific aim could be to produce short (but strong) stalked wheat plants with large seed heads. In order to try and produce these plants the plant breeder has to select the plants which have the required characteristics, plant the seed, cross pollinate them and then collect their seeds.

Figure 5.22

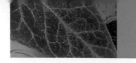

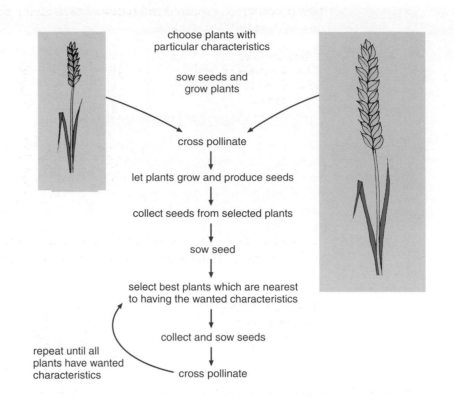

choose plants with
particular characteristics

sow seeds and
grow plants

cross pollinate

let plants grow and produce seeds

collect seeds from selected plants

sow seed

select best plants which are nearest
to having the wanted characteristics

collect and sow seeds

repeat until all
plants have wanted
characteristics

cross pollinate

If the seeds from a short stalk are sown, and the seeds from a tall, large-headed stalk are sown and cross pollinated, when the plants grow to maturity, the best of the crop can be selected and grown. These are again cross pollinated and allowed to make seed.

Figure 5.23
*Wheat in fields
used to consist of
tall stalks with
some short ones The
number of seeds in
the heads of the
wheat also varied*

Again the shortest and largest headed stalks of wheat are cross pollinated. Their seeds are collected and sown.

Figure 5.24
*Variety in wheat
plants*

This procedure is repeated and repeated until all the seeds produce short stalks, carrying big heads of corn.

Figure 5.25
Over a number of generations, the wheat plants become more similar

In a modern field of wheat, the plants are almost identical. All the stalks are uniformly short. All the heads of wheat are big with lots of seeds.

Figure 5.26
Eventually all the wheat plants are almost identical

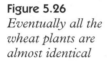

This means that:

- the wheat is less likely to be wind damaged
- the combine harvester can work easily in the field
- there is a higher yield per field
- less plant energy has gone into making a tall stalk and more into making seeds.

Figure 5.27
A field of modern wheat, showing that all the stems are of similar short height with heads of large seeds

Selective breeding in animals

One problem with this process is that most animals take longer to grow to maturity and produce offspring than plants. The second problem is that a single head of wheat will produce perhaps 20 seeds, but a cow, for example, produces only one or two calves each year.

The principle of selective breeding in animals is the same as that for plants – animals with the desired characteristics are cross bred and the offspring observed and perhaps crossed again until the final 'cross' is achieved. Obviously this process takes a long time.

Did you know?

Artificial insemination has meant that sperm can be taken from bulls and put into the best cows much more easily than when the cow or bull had to be taken to a farm!

Did you know?

Australia originally had no wild or domestic cattle. The cattle found there now were brought in by settlers.

The selective breeding programme to produce a drought-resistant breed of cattle in Australia took eight years before enough offspring of several crosses were available. Farmers then chose the variety which best suited the conditions. It was as a result of this that the farmers selected the cross now known as 'Droughtmaster'.

'Shorthorn' bred in Australia from cattle the British settlers took with them, were used to establish the cattle industry in the northern areas of Australia. Their good qualities are that they are fertile, docile, easy to milk and cross breed well. However, they were very susceptible to cattle tick (insect pests).

The Brahman breed of cattle came originally from the 'Bos indicus' in India. These cattle have short thick glossy hair which reflects sunlight, deeply pigmented skin which keeps out the Sun's rays, loose skin which increases the surface area for cooling and sweat glands which allow the body to cool quicker and seem to deter insects!

Cross breeding programmes resulted in the cattle and their descendants being carefully monitored for eight years before being made available to ranchers in 1941. Further studied selection and planned mating fixed the characteristics required to survive in the Northern Territory. In 1956 the breed was named Droughtmaster. This contains approximately 50% Shorthorn and 50% Brahman. The Droughtmaster breed is basically red. They have a high resistance to ticks, they can tolerate heat and drought, they have good fertility, calve easily, have a quiet temperament and are a good beef animal.

It has been found that the Brahman and Shorthorn cattle can both survive well at temperatures down to −13°C and up to 21°C. However temperatures from 21°C up to 24°C cause the European Shorthorn cattle to suffer. The Indian Brahman can tolerate temperatures up to 41°C.

Figure 5.28
Selective breeding in cattle

Variation and selective breeding

Humans have been selectively breeding crops and animals ever since they have been used as sources of food. For example, grasses with desirable characteristics have been crossed repeatedly for thousands of years to produce wheat with ever increasing yields or resistance to disease. Care has to be taken that selective breeding does not end up producing, a limited number of varieties of wheat. If this were to happen then the varieties of wheat are likely to be genetically very similar. Should environmental conditions change in some unforeseen way or some new disease appear, then because of the genetic similarity of the wheat most could be affected as only a few might have the allele that makes them resistant to the changes.

Cloning genes – genetic engineering

Genetic engineering involves altering the genetic make-up of an individual. The principles in all examples and the technique involved are the same – a section of DNA is moved from one species to another to manufacture useful biological products. Using this technique, proteins and drugs can be produced, as well as hormones such as insulin.

The genetic engineering of insulin is described below.

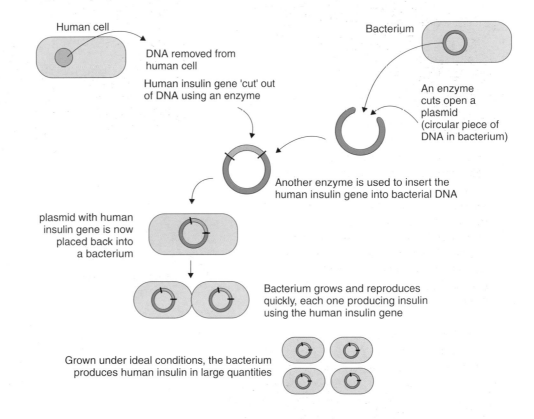

Figure 5.29

Cloning animals

- In 1997 a sheep clone named Dolly was created using the genetic material taken from an udder cell of an adult ewe. This genetic material was joined to an unfertilised egg cell from which the nucleus had been removed. This egg cell was allowed to develop into an early embryo and then was put into the womb of another ewe. Dolly was genetically identical to the ewe whose udder cells were used. In 1999 some scientists who were examining Dolly suggested that she could be showing signs of getting old too early (premature ageing).

- The cloning of pigs by similar methods is more difficult because pigs need several developing embryos in their wombs to maintain pregnancy.

Cloning animals – embryo transplants

Embryo transplantation is one of the latest cloning techniques being attempted in the breeding of mammals including mice, sheep, cattle and horses. Cells are taken from embryos that are only a few days old. At this stage the embryo is made up of a mass of cells that have not yet become specialised into any particular body cell. These cells can be separated, introduced into the wombs of adult females where they can develop normally.

Figure 5.30
Dolly the sheep

Cloning quickly and cheaply produces a large number of genetically identical organisms. Cloning reduces the genetic variation in a species. So if at some time in the future all the sheep in the world were clones from the same parent and conditions changed, all could die as none might have the allele to make them resistant to the change. As with selective breeding, any reduction in the variety of organisms in a particular population decreases the number of different alleles available. This reduces the chances of new varieties being produced, either naturally or artificially, when conditions change and threaten the existence of the population.

Genetic engineering – gene transplants

Genetic engineering means that scientists can now isolate specific genes that carry out a specific function. Such functions can include the production of a particular insecticide or resistance to a weedkiller. These genes can be transferred into plant cells at an early stage of development so that the plant develops with the desired characteristic.

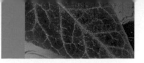

The genetic modification of crops is often carried out in order to produce greatly increased yields as cheaply as possible. One such modification is to make the plants produce an insecticide that kills any insects that try to eat the plants, so reducing the cost of spraying with insecticides. There is concern that this could destroy a wide range of insects, many of which are beneficial in pollination or form part of a food chain. Another concern is when the modification is to make the crop resistant to the action of herbicides. This means that the fields can be sprayed with a weedkiller that will get rid of the weeds but not harm the growing crop. The concern here is that the herbicide-resistant allele could spread into other plants or weeds, perhaps through cross-pollination.

Summary

- By taking cuttings new plants can be produced quickly and cheaply.

- Cuttings will be genetically identical (**clones**) to the parent.

- **Artificial selection (selective breeding)** is used to produce organisms with a particular desirable characteristic.

- Artificial selection and cloning reduce the number of alleles available in a population.

- A reduction in the number of alleles in a population could reduce the chances of producing new varieties if conditions change to threaten a particular population.

- Modern cloning techniques include:
  - tissue culture
  - embryo transplants.

- Genes can also be transferred to the cells of animals so that they develop with desired characteristics. This is called **genetic engineering**.

- The culturing of genetically engineered bacteria on a large scale produces large quantities of useful products e.g. insulin.

- There are economic, social and ethical issues concerning cloning and genetic engineering.

Topic questions

1 Many types of plants can be grown by taking cuttings.

   a) Why do the cuttings from a particular plant all produce the same colour of flower as the parent plant?

   b) Why do the cuttings from a particular plant all produce the same colour of flower as each other?

2 What is selective breeding?

3 What are the advantages of producing plants using a tissue culture technique?

4 Why is important to make sure that a variety of alleles are always available in any particular population?

5 Describe what happens during the production of genetically engineered insulin.

6 Embryo cloning uses embryos that are only a few days old. Why?

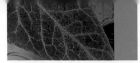

5.5

Co-ordinated	Modular
10.19	Mod 04
	13.3

Evolution

A **species** is a group of organisms which look similar and which can breed together producing **fertile** offspring. Examine the birds below, all of which can be seen in gardens. The species differ not only in size and colour but also in behaviour, call sounds and nest building. These factors keep the species separate.

Figure 5.31
Variety in bird species

What are fossils and how were they formed?

Fossils are the remains or imprints of plants and animals that lived millions of years ago.

Fossilisation

When an organism dies, it is normally decomposed by bacteria and fungi very quickly, leaving nothing to see except possibly a few bones. Fossils are formed where the organism fell into a marsh or bog in which the conditions – no oxygen and water with a low pH – prevented decay. The remains were buried under layers of vegetation and became compressed with age.

Alternatively, fossils can also form where organisms have been covered with layers of sand, volcanic ash or silt. The remains get buried under layers of vegetation and become compressed with age and impregnated with mineral salts from water, turning them into stone. There are other interesting remains such as insects preserved in fossilised tree resin called amber, dinosaurs footprints, petrified trees, remains preserved in ice and coprolites (fossilised droppings).

Fossils can show the time sequence of the existence of, and changes to, organisms over a very long period of time. In sedimentary rocks the material is laid down in layers so the deeper down you go, the older the material and so from top to bottom the fossils are arranged in order of increasing age. These layers can be investigated at cliffs, gravel pits and in spectacular scenery such as the Grand Canyon in Arizona.

The depth of the face is 1700 m and this represents a time span of 500 million years of material.

How fossil evidence supports the theory of evolution

Geologists can trace changes in climate and rock movements and at the same time palaeontologists (fossil experts) can follow changes in the shape and structure of some animals and plants which could result from environmental changes.

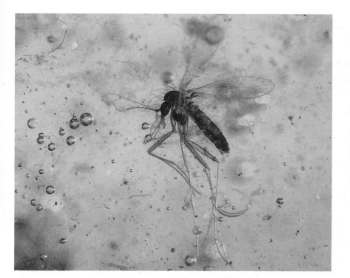

Figure 5.33
The Grand Canyon

Figure 5.32
An insect trapped in amber

The fossil record presents us with the following information.

- There were only a small variety of the simplest marine invertebrates in the bottom layers (oldest rocks) and these represent several million years without much change.

- The presence of fossils in one layer but not in any of the layers above it shows that many forms have become **extinct**.

- There are gaps in the fossil record. There was great excitement when the Archaeopteryx was found, as this is thought to be the 'link' between birds and reptiles. Fossil experts hope that more links will be found.

- There is a definite order in which the fossils appear.

    1 Marine invertebrates – many of these would have been soft bodied and so only the impressions remain. There were many trilobites which are an extinct form of arthropod, some as large as 40 cm. There were also crustaceans, molluscs and primitive starfish. The only plant life was seaweed.

    2 Land plants – the first of these were dependent on water for support and reproduced by means of spores.

    3 Land invertebrates that were air breathing, such as wingless insects, millipedes, spiders and crustaceans that looked like woodlice, have been found in deposits 350 million years old.

    4 The first fish had bodies covered with bony plates; later fossils show scales. Some of these fish have paired limb-like fins that seem to be a transition towards amphibians. Sharks were also around.

    5 Winged invertebrates developed.

    6 As there was more dry land, reptiles evolved. There were both herbivores and carnivores with distinctively-shaped teeth like mammals today. Other reptile groups included turtles and primitive crocodiles. These were followed by the dinosaurs.

    7 Reptiles evolved into birds which had feathers but also reptile-like teeth, solid bones and jointed tails.

    8 Small mammals appeared.

    9 Flowering plants developed.

    10 Development of larger mammals.

Remember, just as today, the majority of plants and animals that died would have been eaten or been decomposed. Only a very few will have died in conditions that would result in fossilisation.

Extinction

An animal or plant is extinct if there are no living specimens. The evidence of these specimens having lived in the past is in the fossil record. The fossil record shows several episodes of mass extinction when many species of plants and animals disappeared, and these have been related to catastrophic changes in climate. Fossil evidence indicates that the dinosaurs became extinct 65 million years ago. A popular explanation for this is that an asteroid or comet struck the Earth causing acid rain and clouds of dust, which blocked out the sunlight leading to global cooling and vegetation changes. Other scientists have suggested massive volcanic activity as the cause. Both ideas leave us with the problem of explaining how the ancestors of dinosaur groups, such as crocodiles, frogs, snakes and insects, survived.

A less dramatic explanation could be that the climate was warmer in the dinosaur age and these large creatures warmed themselves in the Sun whilst browsing on the bushy vegetation. The early herbivorous dinosaurs had large bodies, small brains and, skeletal evidence of their breathing systems suggests, a slow metabolism. The more recent dinosaurs were smaller and had a faster metabolism and these would have been selected against by environmental change, as the climate became cooler and drier. At this time, the active species of insects, reptiles and mammals were increasing and conditions would have been more favourable for these organisms.

Figure 5.34
A reconstruction of an Archaeopteryx

Figure 5.35
A woolly mammoth which became extinct at the time of the last Ice Age

Another much more recent extinction was the woolly mammoth that died out about 10 000 years ago. We know much more about mammoths because some were trapped in ice crevasses and have been preserved as frozen specimens. They had adaptations that allowed them to survive in the cold:

- short dense under-fur with a longer coarse top coat
- skin with an insulating layer of fat and a hump-like fat reserve on the back
- deep scratch marks on the lower side of the huge tusks indicated that they could be used for clearing snow from the grass for feeding
- short ears, tail and legs to reduce heat loss.

Although mammoths were hunted, it is thought that they also became extinct as a result of selection against them after a climate change because there was a reduction in grassland at the time of the Ice Age.

Extinction may be caused by:

- changes in the environment
- successful *new* predators
- successful *new* competitors
- *new* diseases.

Unless evolution occurs so that species become better adapted to these changes, they may become extinct.

The rise of antibiotic-resistant bacteria

Some microorganisms produce substances that prevent bacteria from reproducing. Such substances, which include penicillin, are called antibiotics. Antibiotics have no effect on the action of viruses. The ability of penicillin to prevent the reproduction of bacteria was first observed in 1928. It was not used on humans until 1940 when it was used to fight bacterial infection in wounded soldiers. For the next few years penicillin was considered to be a 'wonder-drug' as it was able to control almost every bacterial infection. In the 1950s strains of bacteria were appearing that were unaffected by penicillin. This was not considered a problem as more and more varieties of antibiotics were produced. Some of the main problems during this time were that:

- too many patients thought that antibiotics were the cure for any illness, including those caused by a virus

- too many doctors were willing to prescribe antibiotics, often inappropriately

- farmers and vets were using large quantities of antibiotics on food-producing animals.

It was this over-use of antibiotics that has resulted in a large majority of the disease-causing bacteria now being unaffected by the action of the majority of antibiotics.

How do bacteria become resistant?

Bacteria can, in an ideal environment reproduce asexually every 20 minutes. This rapid rate of reproduction can produce changes in the genetic material as the DNA tries to make copies of itself during mitosis (see section 1.5). In some instances, the mutation in the DNA results in a change to an enzyme structure in the bacterium that makes the bacterium resistant to the action of the antibiotic. This resistant bacterium will reproduce and very quickly form clones of antibiotic-resistant bacteria.

Charles Darwin and his theory of evolution

Charles Darwin (1809–1882) was a well-travelled naturalist and a breeder of fancy pigeons. He kept very careful records of the parents and offspring of his pigeon crosses. He made a 5 year voyage to South America and the Galapagos Islands from 1831 to 1836, during which he kept very careful observations. These observations were the basis of the paper he presented with another scientist called Alfred Wallace to the Linnean Society in London in 1859. This paper was called 'On the Origin of Species by Means of Natural Selection'. At the time that Darwin proposed the ideas of **natural selection**, there was no knowledge of genetics and his theory was based on breeding results and observations only.

He made the following propositions:

- All organisms produce more young than the environment can easily support – there is often a shortage of food.

- Variations exist within populations and few individuals are exactly alike in any measurable variable but some of the features seen in parents are also seen in their offspring.

- There is a struggle for existence – those organisms best suited to the environment survive to reproduce and have more offspring than those less well adapted and therefore the beneficial characteristics become more common. (The reverse argument can be used to explain the extinct forms found in the fossil record only.)

- The result of the 'struggle for existence' and limited food and shelter sites in the environment means that populations remain approximately constant in a balanced ecosystem until they are affected by man or natural disasters.

Reactions to Darwin's theory of evolution

When Darwin published his theory of evolution in 1859 there was a great deal of opposition to the ideas contained in it. Most opposition came from those people who considered that Darwin's ideas were a serious challenge to the religious teachings about the story of creation. Even many scientists criticised the ideas because:

- there was no laboratory proof so it must remain a theory

- natural selection did not explain the big changes shown by fossil records

- there was no explanation as to how variations within a species arise, nor how they were passed on to future generations.

Gradually during the rest of his lifetime Darwin's ideas became more acceptable. But it was not until early in the 20th century, when Mendel's work on genetics (see section 5.1) was used to help explain the process of natural selection, that Darwin's ideas were almost universally accepted.

> **Did you know?**
>
> The popular idea that human beings were descended from apes was not part of Darwin's theory. What he did say was that human beings, apes and monkeys all descended from the same primitive ancestors.

How Mendel's work helps to explain natural selection

Knowledge of genetics enables some of the gaps in Darwin's theory of evolution to be explained.

- Variations in the population can be explained by spontaneous mutation that may not be caused by environmental factors. Unless these mutations are lethal, they will be passed on to the offspring.

- As a result of the formation of gametes and fertilisation in sexual reproduction, offspring will receive a selection of alleles from both parents and will have a different combination from their parents. (They may have Mum's nose and build but Dad's height and shape of fingers.)

- Some combinations of alleles are better suited to particular environments and have a selective advantage. Individuals with such combinations will survive and breed successfully.

- Over many generations, natural selection results in an increase in the spread of beneficial mutations and a reduction in that of harmful mutations.

Conflicting theories of evolution

The fossil record provides evidence of change over very long periods of time and the development of antibiotic resistance in bacteria gives evidence of changes still taking place in organisms. Darwin explained these changes through his theory of evolution, however a scientist, called Lamarck, had a different explanation.

Jean Baptiste Lamarck (1744–1829) was a French natural philosopher. Lamarck's theory was related to people and was meant to please the politicians. He suggested that a desire or need for change caused that change to happen in the organism and that this would be passed on to the offspring. Structures that are constantly in use are well developed and characteristics acquired during life would be passed on. For example, the blacksmith needed big muscles and would pass these on to his son. His 'evidence' was that when the environment causes the need for a structure, this induces the structure that will help the organism adjust to the environment.

His famous example was the giraffe. Lamarck suggested that the giraffe acquired its high shoulders and long neck by straining to reach the leaves in the tree, and that these characteristics were passed on to the offspring which by continual stretching extended them more. Gradually, generation by generation, giraffes got longer necks.

How Darwin's theory of natural selection accounts for the long necks

- Although most of the ancestors of giraffes had short necks there would be some with longer necks – there was variation and the characteristics were inherited.

- If the food supply became in short supply – a struggle for existence would occur.

- Those animals best able to find food would survive. In this case those with long necks could reach leaves higher up in the trees. These would survive and breed offspring with long necks – survival of the fittest.

- Nature has selected those most suited to the changing conditions – natural selection.

- If this continues over very many generations the present day giraffe with its very long neck would evolve – evolution.

Summary

◆ **Fossils** are the evidence of plants or animals from many millions of years ago.

◆ Fossils provide evidence of how organisms have changed.

◆ The theory of evolution states that all species of living things alive today – and many others that have become **extinct** – have evolved from simple life forms which first developed more than three billion years ago.

◆ Evolution occurs through **natural selection** in which:

- individuals within a species may show variation because of genetic differences
- predation, disease and competition cause large numbers of individuals to die
- individuals with characteristics most suited to the environment are more likely to survive and breed successfully
- the genes which have enabled these individuals to survive are then passed on to the next generation.

◆ It is possible to explain:

- how fossils provide evidence for the theory of evolution
- how the over-use of antibiotics can lead to the evolution of resistant bacteria.

◆ The conditions in which a species survives may change. Unless evolution occurs and the species become better adapted to survive the changing conditions they may become extinct.

◆ It is possible to suggest why Darwin's theory of evolution was only gradually accepted.

◆ There are differences between Darwin's theory of evolution and those of Lamarck.

Topic questions

1 a) What are fossils?
   b) Give four different ways fossils can be formed.

2 What does the fossil record show?

3 Why do some organisms become extinct?

4 What are the key points in Darwin's theory of evolution?

5 What was Lamarck's theory of evolution?

Examination questions

1 The drawing shows some of the stages of reproduction in horses.

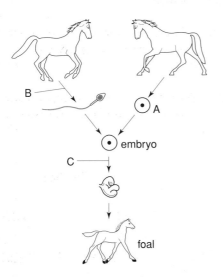

a) i) Name this type of reproduction. *(1 mark)*
   ii) Name the type of cell labelled **A**. *(1 mark)*
b) Name the type of cell division taking place at the stage labelled:
   i) **B**
   ii) **C**. *(2 marks)*
c) How does the number of chromosomes in each cell of the embryo compare with the number of chromosomes in cell **A**? *(1 mark)*
d) When the foal grows up it will look similar to its parents but it will **not** be identical to either parent.
   i) Explain why it will look similar to its parents. *(1 mark)*
   ii) Explain why it will **not** be identical to either of its parents. *(2 marks)*

2 This couple has just found out that the woman is pregnant. They wonder whether the child will be a boy or a girl.

Sex chromosomes	Sex chromosomes ....

a) Fill in the boxes to show the sex chromosomes of the woman and the man. *(2 marks)*
b) The couple already has one girl. What is the chance that the new baby will be another girl? Explain the reason for your answer. You may use a genetic diagram if you wish. *(3 marks)*

3 In the 1850s an Austrian monk, called Gregor Mendel, carried out a series of investigations on heredity.

a) i) What plants did he use for his investigations? *(1 mark)*

   ii) In his work he assumed that one gene controlled one characteristic. He started his investigations with pure breeding parents. Use a genetic diagram to show how he explained the following result.
(4 marks)

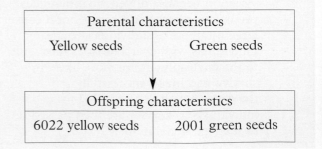

Parental characteristics	
Yellow seeds	Green seeds

Offspring characteristics	
6022 yellow seeds	2001 green seeds

4 Giraffes feed on the leaves of trees and other plants in areas of Africa. They are adapted, through evolution, to survive in their environment.

a) Use the information in the picture to give one way in which the giraffe is adapted to its environment. *(1 mark)*

b) Explain how Jean-Baptiste Lamarck (1744–1829) accounted for the evolution of the long neck in giraffes. *(3 marks)*

c) Another scientist, August Weismann (1834–1914) wanted to check Lamarck's explanation. To do this he cut off the tails of a number of generations of mice and looked at the offspring. His results did not support Lamarck's theory. Explain why. *(2 marks)*

d) Explain how Charles Darwin (1809–1882) accounted for the evolution of the long neck in giraffes. *(4 marks)*

5 Sickle cell anaemia is an example of a disease caused by a *mutation* affecting one of the genes involved in the production of haemoglobin.

- Hb is a gene that determines haemoglobin.
- Hb^A causes normal haemoglobin and is *dominant*.
- Hb^S causes defective haemoglobin and is *recessive*.
- In the *homozygous* recessive condition the person suffers acute anaemia and has a low life expectancy.
- In the *heterozygous* condition individuals suffer from the sickle cell trait but have increased resistance to malaria.

a) What is the role of haemoglobin in the body? *(1 mark)*

b) What is a *mutation*? *(1 mark)*

c) Use the information above to explain what is meant by the terms *homozygous* and *heterozygous*. *(2 marks)*

d) Use the information above to explain what is meant by the terms *dominant* and *recessive*. *(2 marks)*

e)

		Father	
		Hb^A	Hb^S
Mother	Hb^A	**Child 1** $Hb^A Hb^A$	**Child 2** $Hb^A Hb^S$
	Hb^S	**Child 3** $Hb^A Hb^S$	**Child 4** $Hb^S Hb^S$

i) Which child will have sickle cell anaemia?
ii) Which child will have sickle cell trait?
iii) Which child will not carry any sickle cell genes?
iv) Which child will be more resistant to malaria? *(4 marks)*

Chapter 6
Living things in their environment

Key terms   adaptation · biomass · competition · consumer · decay · decomposer · deforestation · detritivores · ecosystem · environment · eutrophication · fertiliser · food chain · global warming · greenhouse effect · herbicide · herbivore · migration · nitrifying bacteria · organic · pesticide · pollution · population · predation · predator · prey · producer · putrefying bacteria · pyramids · resources · toxic

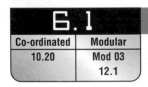

6.1   Adaptation and competition

Co-ordinated	Modular
10.20	Mod 03 12.1

All living things have to be able to cope with the changes going on in the environment around them. Living things reproduce within their **environment** and this creates a **population** of a particular organism which compete for the food, water and space within the area.

As the population of an organism grows, the **competition** for these **resources** increases until only those which are able to compete best can survive. A typical growth curve for a population is shown below.

Figure 6.1
A typical population curve

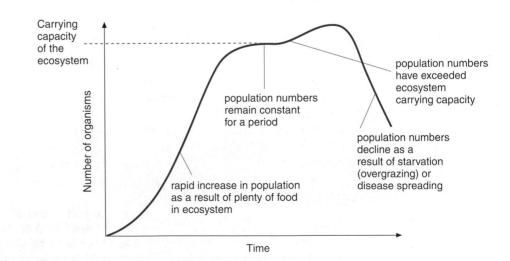

In the early stages the population grows rapidly because there is plenty of food, water and space available. As numbers increase, the food supply may begin to run out, some of the organisms may starve and the growth in the population begins to slow down.

110

Eventually a point is reached where as many organisms are dying as are being produced. The population becomes stable and remains so as long as the conditions of food, water and space also remain constant.

In some areas, animals may migrate to avoid overpopulating an area. As the population increases, the pressure on the food supply also increases. Animals may move around an area or move to a completely different area to ensure that the food supply is not completely exhausted or to give it time to recover. In places where there is a marked seasonal change – hot/cold or wet/dry – the **migration** is an annual occurrence.

Distribution and abundance of organisms

What happens in nature, however, is not always as simple as the situation described in Figure 6.1. There are often other forces that control populations, such as **predation** – one animal preying on another. The **predator** may reduce the population of its prey until there is not enough for it to feed on. The predator may then begin to starve. At this point, the **prey** population may rise again because there are not enough predators to control numbers. As the population of the prey rises, the predators again have enough to eat and their population grows. This pattern is often repeated over many generations.

The diagram below showing the relationship of the vole (prey) and the fox (predator) clearly demonstrates these changes.

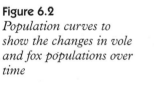

Figure 6.2
Population curves to show the changes in vole and fox populations over time

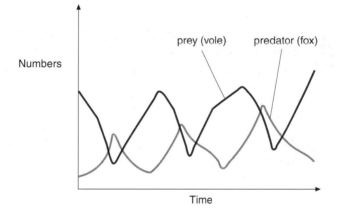

So the size of a population may be affected by:

- the total amount of food or nutrients available
- the total amount of water available
- competition for food or nutrients
- competition for water
- competition for light
- predation or grazing
- disease.

Adaptations for survival

Each type of environment poses problems to the plants and animals living in it. Those organisms that become adapted to the environment are the ones that will survive. Those that cannot adapt often become extinct. The arctic and desert environments represent two extremes in one sense but they are similar in another. One is very hot and the other very cold. In both, however, water is in short supply.

In the arctic it is very cold, much too cold for most organisms. Only those which are protected against the cold or can conserve heat can survive.

a)

b)

Figure 6.3
*a) An arctic fox, b) a
fennec (desert) fox*

Adaptations of organisms to the cold environment

The arctic fox is an animal that can survive these extreme conditions. Its adaptations for survival are given below.

- Thick fur, which insulates the body.

- The fur is white – this gives the fox camouflage so that it stands a better chance of catching prey. The white fur also reduces the rate at which heat is lost by radiation.

- Mammals like the arctic fox lose heat most easily through ears and feet. The ears of the fox are very small so their surface area is also very small in relation to the volume. The bigger the surface area in relation to the volume, the more likely the ears are to radiate heat from the body.

- The feet, which are in contact with the ice, are also very small and cold. This prevents heat being lost from the feet by conduction into the ice.

Animals such as the polar bear have a thick layer of body fat under the fur. Fat is a good insulator and also provides a lot of heat when it is used as a source of energy in respiration. Whales living in the arctic waters cannot have a thick layer of fur as it would slow them down in the water. They have a thick layer of fat called blubber instead.

Adaptations of organisms to the desert environment

The arctic and the desert are similar in that water is hard to come by. In the arctic it is frozen; in the desert there is little water either because it rains very little or because when it does rain the water evaporates so quickly from the surface – the only water is often deep underground. Deserts are usually hot places so the animals are faced with the problem of trying to lose heat rather than retain it.

One animal which has adapted to these conditions is the fennec fox. Its adaptations are given below.

- Large ears in relation to its body – the ears act as radiators, releasing heat.
- The fur is very thin so that more heat can be lost from the body.
- There is little body fat.
- The fox is a brown colour to camouflage it against the sand.

Plants too have to survive in harsh conditions. Cacti are plants which live in deserts, often where it may not rain for many years. They have adapted to survive these conditions as shown below.

Figure 6.4
*Cacti are well adapted
to survive the harsh
desert conditions*

- The stems of the cacti are thick and fleshy for storing water.

- Often the leaves have been replaced by spines which do not lose water as rapidly as leaves; spines also prevent animals from eating the stem to get at the water it contains.

- The cactus stems are green because they contain chlorophyll for photosynthesis.

- The roots of the cactus are very long so it can collect water from deep in the soil over a wide area.

Adaptations of plants to crowded conditions

Plants have to be able to compete for light and space when they grow in crowded conditions. The tallest plants are often the ones that survive. In deciduous woodland, the tall trees prevent much of the light reaching the ground, but there are often many plants living at ground level. These plants produce their leaves in early spring before the leaves on the larger trees appear. They flower and then remain dormant in the soil until the following spring. The energy they have stored remains

in the part of the plant below ground as a bulb or corm. The decaying leaves that fall from the trees in the autumn provide the ground-living plants with a rich source of nutrients.

Figure 6.5
The bluebells flower and set seed in spring before leaves of the trees grow and shade them

Summary

- Organisms have features that help them survive in their own environment. For example, some animals and plants are **adapted** for survival in arctic environments.

- An environment has a limited amount of resources which organisms must compete for.

- Animals which kill and eat other animals are **predators**; the animals they eat are called **prey**.

- In a community
  - if the population of prey increases more food is available for its predators and their population may rise.

  - if the population of predators increases more food is needed and the population of prey will decrease.

- The size of a population may be affected by
  - the total amount of food or nutrients available
  - the total amount of water available
  - **competition** for food or nutrients
  - competition for light
  - competition for water
  - predation or grazing
  - disease.

Topic questions

1   a) What is meant by a population?
    b) What factors control the population size of a particular animal?

2   a) Foxes eat rabbits. Which is the predator?
    b) What happens to the populations of predators and their prey over a long period of time?

3   a) Why are animals that live in very warm climates likely to have large ears?
    b) In what ways are cacti adapted to live in very dry areas? State how each adaptation helps the plant to suvive.

6.2	Human impact on the environment

Co-ordinated	Modular
10.21	Mod 03 12.4

Pollution from fossil fuels

All organisms make an impact on the environment. In many cases it is very small. Humans, however, have made a huge impact on the environment. Some of this has been beneficial to the planet but other activities have caused problems, especially for other organisms.

Since the Industrial Revolution there has been an increasing reliance on the use of fossil fuels as a source of energy. Fossil fuels were formed many millions of years ago from the remains of dead micro-organisms and plants. Normally the organisms would have decayed and they would have been returned to the environment for recycling as carbon dioxide, but the organisms did *not* decay and the carbon compounds became locked in the soil. The micro-organisms became oil and natural gas and the plants became coal.

When the fossil fuels are burned, the chemicals they contain are released into the atmosphere. One of the gases formed in the process of combustion is carbon dioxide. This is a gas which is found naturally in the atmosphere. It is vital for photosynthesis, however there is evidence that the amount of carbon dioxide in the atmosphere is increasing.

There are also compounds of sulphur and nitrogen in fossil fuels. When fossil fuels are burned they release sulphur dioxide and nitrogen oxides, both of which are poisonous gases. Sulphur dioxide and nitrogen oxides are strongly acidic. These gases are soluble and when they dissolve in rain they make the rain acidic. This 'acid rain' falls on soil where it causes mineral salts in the soil vital to plant growth to be removed. The acid rain also has a direct effect on the plants by damaging roots and leaves. The rain eventually finds its way into rivers, streams and lakes, making the water too acidic for organisms to live.

Acid rain can also attack the stonework of buildings causing it to dissolve and crumble.

Figure 6.6
Acid rain has severely damaged these trees

The increase in the levels of carbon dioxide in the atmosphere is causing concern in another way. Carbon dioxide absorbs infra-red light (heat rays) from the Sun. If the amount of carbon dioxide increases, more heat will be absorbed and become trapped in the upper atmosphere. The atmosphere will behave exactly like the inside of a greenhouse and it will heat up (the '**greenhouse effect**'). The effect of this will be to cause the temperature of the whole Earth to increase. Possible effects of this '**global warming**' might include:

- the melting of the ice caps at the poles, increasing sea levels and causing widespread flooding of low-lying coastal regions.

- climatic changes which might alter the wind and rainfall patterns and change the distribution of plants and animals on the Earth, perhaps increasing the numbers of pests and the incidence of serious diseases.

Deforestation

As the human population has grown, it has demanded more and more living area, space for growing crops and raw materials. These demands have often been met by cutting down forests. The space created by this is used to grow crops, build homes

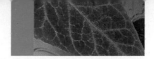

and roads (this is called urban development) and the trees have been used to make furniture and other items. As long as the cutting down of the forests is carried out at the same rate as growth, there is no harm done, but recently the rate of **deforestation** has greatly exceeded the rate of growth of the forests (reforestation).

The areas used for agriculture often yield only one or two crops before the thin soil is exhausted and has run out of nutrients. The areas used for urban development are lost forever.

Less trees also means less carbon dioxide removed from the atmosphere for photosynthesis and so deforestation can contribute to global warming.

Problems resulting from deforestation as seen in Brazil

The vast rainforest of Brazil had its own climate but the removal of huge areas of trees has changed this.

- The overall rainfall has reduced, making the climate drier but now the rain often falls in heavy short bursts.

- The canopy of trees once prevented the rain from hitting the soil but now the trees have gone, the heavy rain falls straight on the soil washing it away.

- The soil has been eroded and cannot support the growth of crops. The crops fail and the land is abandoned.

- The removal of the trees has had a terrible effect on the wildlife in the area. The number of different species that can be supported by the dwindling vegetation has been greatly reduced.

- Reduction in species diversity has a destructive effect on food chains and webs and has led to the extinction of many species of plants and animals.

Greenhouse effect and global warming

Carbon dioxide is not the only gas responsible for the **greenhouse effect**. Methane, though much less common in the atmosphere than carbon dioxide, has a greater 'greenhouse effect'. A certain volume of methane has about 20 times the 'warming potential' than the same volume of carbon dioxide. Like carbon dioxide, it too is increasing due to human activities.

Where does the methane come from?

Much of the methane released into the atmosphere comes from rice fields and from the digestive system of cattle.

Did you know?

Although the amount of methane in the atmosphere is low, about 20% of it is produced from rice paddies and cattle.

Methane and rice paddies
The flooding of a rice field encourages the anaerobic fermentation of organic soil matter. This fermentation produces large amounts of methane that is released through the roots and stems of the growing rice plants.

Methane and cattle
The breakdown of grass in the digestive system of cattle (and other animals that 'chew the cud') produces large amounts of methane which the animal releases into the atmosphere.

Figure 6.7
Flooded rice paddy field

Food production

There is no doubt that humans have been very successful at increasing the amount of food produced by agriculture. Much of this has been achieved by increased efficiency through:

1   the elimination of pests which destroy food crops or compete with farm animals for food supplies

2   the elimination of weeds which compete with plant crops for water, light and nutrients

3   the use of **fertilisers** to make crops grow faster and give bigger yields

4   the use of larger fields so that more crops can be grown in a given area.

Farmers have a duty of care, however. This means that while they try to increase yields and eliminate pests, they must also be aware of the effects on the whole agricultural environment caused by pesticides, fertilisers and increase in field sizes by removal of hedgerows.

Sadly in the past some farmers have neglected this duty of care.

Damaging ecosystems

Pesticide use

The pesticides used have, in many cases, turned out to be toxic to other animals which might be helpful to the farmer or they have disrupted food chains causing the starvation of animals further up the food chain. Some pesticides will also kill pollinating species and reduce crop yield.

Excessive use of fertilisers

The use of fertilisers is often not well controlled. Farmyard manure is a very useful fertiliser but its quality cannot be guaranteed. It may not contain enough of the right kind of minerals or too much of another. Although farmyard manure does improve the water retention of the soil, its action is slow because bacteria and fungi have to break it down into the simpler substances the crops needs. Many farmers have resorted to the use of chemical fertilisers because the composition is known and the action is much faster. The problem is that they had little idea of how much fertiliser to apply and as a result they apply too much. The fertiliser which is not taken up by the crop plant is often leached out of the soil by rain into rivers.

The fertiliser causes a rapid increase in the growth of plants in the rivers and streams. The plants often grow so quickly that they completely block out the light for plants below the surface and prevent them producing oxygen by photosynthesis.

The plants die and start to decay. The aerobic bacteria and fungi which decompose the plants use up the oxygen from the water. The amount of oxygen in the water falls until there is not enough for larger animals. These also die and decay, using up more oxygen. Eventually the river or stream becomes devoid of oxygen and the water becomes lifeless.

The addition of excess plant nutrients which results in excess plant growth, death, decomposition of the plants by aerobic bacteria and the removal of oxygen from rivers and streams is termed eutrophication. It can be a serious problem in rivers and streams that flow through agricultural land.

Figure 6.8
Section through a stream showing the effects of eutrophication, causing light to be blocked out from the submerged plants

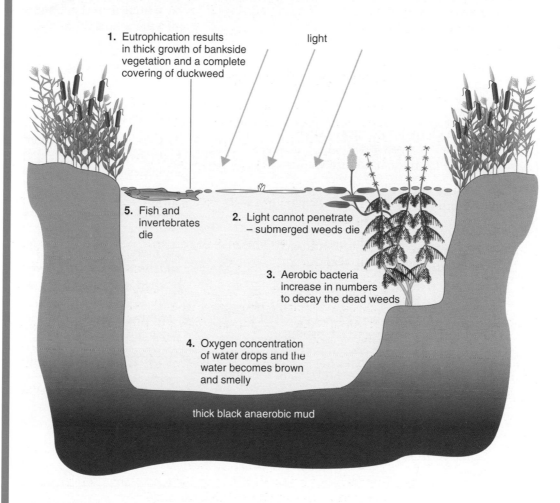

1. Eutrophication results in thick growth of bankside vegetation and a complete covering of duckweed

light

5. Fish and invertebrates die

2. Light cannot penetrate – submerged weeds die

3. Aerobic bacteria increase in numbers to decay the dead weeds

4. Oxygen concentration of water drops and the water becomes brown and smelly

thick black anaerobic mud

The dumping of raw sewage into a lake or river causes the same eutrophication effects.

Making decisions

The solutions to the problems caused by human conflict with the environment will not be easy to solve. There are many factors to be taken into consideration. The human population is rising rapidly and however much care is taken, the conflict with the environment is bound to increase. Humans are faced with a huge extra demand for food and without the use of chemicals, such as fertilisers, it is difficult to see how that demand can be met.

We are faced with a population requiring more leisure pursuits, and dwindling unspoiled areas of land and water. It is vital that these should be managed properly otherwise we could all be the losers as the number of species of animals and plants declines.

We are faced with the prospect of an atmosphere that is changing due to the massive amounts of pollutants being pumped into it. The increases in gases such as methane and carbon dioxide must be slowed down or reversed to prevent global warming becoming a more serious problem than it is now. All the decisions that have to be made will be complex. There will be winners and losers if the right ones are taken, but only losers if the decisions are wrong ones.

Summary

- Humans reduce the amount of land available for other animals and plants by building, quarrying, farming and dumping waste.

- Human activity may **pollute**:
  - water with sewage, **fertiliser** or **toxic** chemicals
  - air, with smoke and gases such as carbon dioxide
  - land, with toxic chemicals, such as **pesticides** and **herbicides**, which may get washed from the land into water.

- The burning of fossil fuels releases carbon dioxide into the air.

- The burning of fossil fuels may also release sulphur dioxide and oxides of nitrogen. These dissolve in rain to make it acidic.

- Rapid growth in the human population and an increase in the standard of living means that:
  - raw materials are being used up rapidly
  - large volumes of waste are produced
  - there is a rapid increase in pollution.

- Large scale **deforestation** for timber and for land for farming has:
  - increased the release of carbon dioxide into the air, because of burning
  - reduced the rate at which carbon dioxide is removed from the air and 'locked up' as wood.

- Increases in the number of cattle and rice fields have increased the amount of methane released into the atmosphere.

- Carbon dioxide and methane in the air absorb much of the energy radiated by the Earth. Some of this energy is re-radiated back to the Earth and keeps the Earth warmer than it otherwise might be.

- The concentrations of carbon dioxide and methane in the air are increasing. These increases enhance the '**greenhouse effect**'.

- An increase of only a few degree Celsius (**global warming**) may cause large changes in the Earth's climate and a rise in sea level.

- Farmers add fertilisers to replace nutrients which crops remove. Excess fertiliser may be washed into lakes and rivers.

- Pollution of water by fertilisers may cause **eutrophication**.

- Untreated sewage provides food for microorganisms. This in water has the same effect (eutrophication) as does dead vegetation.

Topic questions

1 When fossil fuels are burnt, carbon dioxide, sulphur dioxide and oxides of nitrogen are released into the air.
   a) Which of these gases is a 'greenhouse gas'?
   b) Which gases form acid rain?

2 In many parts of the world more trees are being cut down than are being planted. How will the fewer number of trees affect:
   a) the amount of carbon dioxide in the air? Give a reason for your answer.
   b) the amount of oxygen in the air? Give a reason for your answer.

3 Cattle and rice fields produce large amounts of another 'greenhouse gas'. What is its name?

4 Give two possible effects of global warming.

5 Explain how the use of too much fertiliser can cause eutrophication.

6 Explain how global warming is thought to occur.

6.3	
Co-ordinated	**Modular**
10.22	Mod 03
	12.2

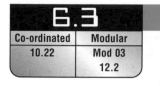

Energy and nutrient transfers

Food chains and webs

All living things need energy. The law of conservation of energy states that energy cannot be created or destroyed. This means that the same amount of energy comes out of an **ecosystem** as goes into it.

An ecosystem is a unit made up of living components – plants and animals – and non-living components, for example a woodland with trees, animals and non-living components such as soil. There is a complex and often delicate balance between the organisms living in an area and the environment.

The primary source of energy for nearly all ecosystems is the Sun because it provides the energy needed for photosynthesis. The plants that can carry out photosynthesis are known as the **producers** because they produce all the food for themselves and are often eaten by other organisms. The organisms that feed on plants are called the primary **consumers**. The primary consumers are fed upon by the secondary consumers and these in turn may be fed upon by the tertiary consumers.

Food chains and pyramids of biomass

Food chains show the feeding relationships and direction of energy transfer between a number of organisms.

It is possible to produce pyramids by measuring the **biomass** in a food chain. These pyramids are usually based on the dry mass of the organisms in a food chain. A pyramid of biomass for the food chain would show a typical pyramid shape.

one sparrowhawk

several bluetits

thousands of greenfly

leaves of one oak tree

Figure 6.9
A pyramid of biomass for an oak tree

oak tree → insects → blue tits → sparrow hawk

The change in the size of the blocks from the base to the tip of a pyramid of biomass shows clearly the energy loss that occurs at each step.

A pyramid of biomass provides a quantitative display of biomass in a food chain. The simple link type of food chain does not show this.

Energy losses along a food chain

At every link in all food chains there is a loss of energy.

All the organisms in a food chain must respire to live. The respiration uses energy that cannot be passed to the next step in the food chain. The energy from respiration is used for movement, growth and heat production. These losses are especially large in mammals and birds. This is because their body temperatures are kept constant, at a level much higher than the temperature of the environment. This requires a lot of energy. None of this energy can be used by the organisms in the next link in the food chain. Eventually there is so little energy left there is not enough to go round to all the organisms in the population.

Notice in Figure 6.10 that the direction of energy flow is always away from the producer to the consumer.

The biggest loss of energy in any food chain is between the Sun and the producer. The producers trap only about 1% of the Sun's energy. The loss in the other steps in the food chain may be as much as 90% in each step.

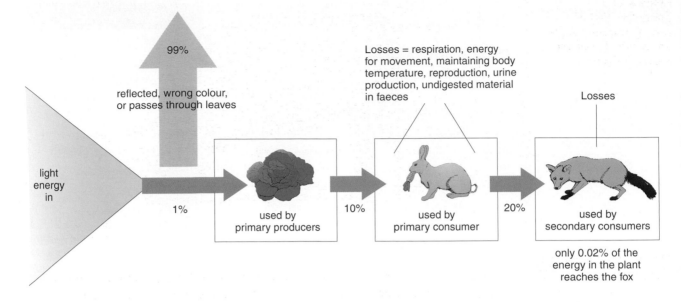

99%

reflected, wrong colour, or passes through leaves

Losses = respiration, energy for movement, maintaining body temperature, reproduction, urine production, undigested material in faeces

Losses

light energy in

1%

used by primary producers

10%

used by primary consumer

20%

used by secondary consumers

only 0.02% of the energy in the plant reaches the fox

Figure 6.10 ▲
A simple food chain showing where the energy is lost

Producing food for humans

There are many food chains which involve humans. In some of these humans are primary consumers.

In other food chains most humans are secondary consumers because most humans eat primary consumers such as cattle (Figure 6.11).

Because energy is lost at each link in a food chain, the fewer steps the food chain has the more energy reaches humans. It should be more efficient for humans to eat plants but unfortunately not all the parts of a plant can be digested by humans and so quite a lot of the energy is lost. It does show, however, that the same area of agricultural land can be used more effectively if used to grow crops than if it is used for the grazing of animals.

Problems involved with the large scale production of food

The cow in Figure 6.12 has transferred only 4% of the energy originally available in the grass into new tissue. If farmers wanted to increase this value then methods would need to be taken to decrease the energy losses due to the animal moving around, needing to keep warm and producing waste. Such methods could include:

● limiting how much the animal moves
● keeping the animal in warm surroundings, especially in cold weather
● controlling the diet to reduce the amount of waste produced.

Figure 6.11 ▶
In the food chain, wheat → pigs → humans, humans are the secondary consumers. The humans receive 20% of the energy from the chain – the remaining 80% is used by the pigs to keep warm and move, or is lost in their waste products

Figure 6.12
For every 100J of energy available in the grass, only 4J gets into the cow's tissues

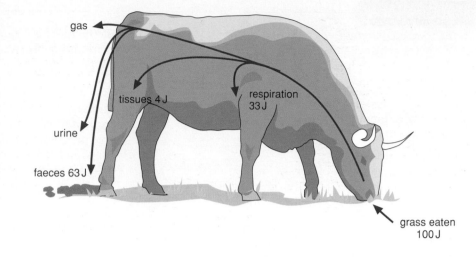

gas

tissues 4 J

respiration 33 J

urine

faeces 63 J

grass eaten 100 J

Much of our fruit comes from other parts of the world. It needs to be harvested, packaged and transported. The energy and resources used to package and transport foods over long distances makes food production inefficient. During each of these processes the quality can deteriorate especially through the action of decomposers. One technique used to stop fruit from decaying before it can be sold is to use special hormones that slow down the rate at which it ripens.

Many of the methods used to improve the efficiency of food production led to the practice of 'factory farming'. Many people consider the confining of animals, especially chickens, in a restricted space and feeding them a specially prepared diet to be cruel. Only recently has there been pressure on farmers to improve the conditions under which the animals are reared. This pressure has been triggered by increased outbreaks of food-poisoning from the eating of eggs and the concerns over the links between cattle food, BSE and CJD in humans.

Summary

- Radiation from the Sun is the source of energy for all organisms.

- Green plants use some of the solar energy that reaches them. This energy is stored in the substances from which the plant is made.

- The mass of living material (**biomass**) at each stage in a **food chain** is less than it was at the previous stage.

- The biomass at each stage can be drawn to scale and shown as a pyramid of biomass.

- Because at each stage in a food chain, less material and less energy are contained the biomass of the organisms, the efficiency of food production can be improved by reducing the number of stages in the food chain.

- The amounts of material and energy in the biomass of organisms is reduced at each successive stage in a food chain because:
  - some materials and energy are always lost in the organisms waste
  - respiration supplies the energy needs for living processes. Much of this energy is eventually lost as heat to the surroundings. These losses are large in warm-blooded animals.

- The efficiency of food production can be improved by:
  - restricting energy losses from food animals by limiting their movement and controlling the temperature of their surroundings
  - using hormones to regulate the ripening of fruits.

- There are positive and negative effects of managing food production and distribution.

Topic questions

1 In the food chain: grass → cattle → humans
   a) Which organisms are consumers?
   b) Which is the producer?
   c) Which is a primary consumer?
   d) Which is a secondary consumer?

2 What is the original source of energy for nearly all living organisms?

3 a) What is biomass?
   b) What is a pyramid of biomass?
   c) What happens to the biomass at each stage in a food chain? Give a reason for your answer.

4 It would be more efficient if humans obtained their energy from directly eating grass rather than from the eating of cattle. Why?

5 a) Why are heat losses greater from cattle and chickens than from salmon and other fish?
   b) Explain three ways in which the efficiency of beef production can be increased.
   c) How can fruit be prevented from over-ripening before it reaches the shops?

6.4

Co-ordinated	Modular
10.23	Mod 03
	12.3

Nutrient cycles

Decomposers and their role in nutrient cycling

When living things grow, they must remove materials from the environment. During photosynthesis, plants take in carbon dioxide from the air and water from the soil and use these inorganic materials to make their food – organic compounds such as glucose and starch. They also take in mineral salts from the soil which help the plants to make other **organic** compounds called proteins.

Note: An inorganic substance generally does not contain carbon; organic substances are carbon-containing chemicals.

These organic compounds are transferred from plants to animals, and from one animal to another along food chains. As plants and animals live, some of the materials are returned to the environment as waste compounds such as carbon dioxide and urea. The process of **decay**, which begins as soon as plants and animals die, helps the return of all nutrients to the environment. The materials are firstly eaten by **detrivores** (detritus-eating species) such as worms, woodlice and maggots and then broken down by decomposers. The **decomposers**, which are bacteria and fungi, are most important in the decay process. They feed on the dead animals and plants or on their waste, turning it back into carbon dioxide, water, nitrogen-containing compounds and mineral salts.

The decay process is vital in the recycling of these materials and allows them to be taken up again by plants. If all the materials were not eventually returned to the environment, some would run out. This can happen when plants are removed, as in deforestation of rainforest areas.

The process of decay can be very slow but it can be speeded up by making sure that the decomposers have warm, moist conditions and a plentiful supply of oxygen. These are the conditions found in a compost heap or sewage farm.

Did you know?

About 300 million years ago, the organic materials did not decay but became coal and oil. We use the chemical energy locked away in coal and oil when we burn them.

Carbon cycle

One important element recycled by the process of decay is carbon. All plants and animals need a supply of carbon. Plants make glucose, during photosynthesis, from carbon dioxide in the air and use the glucose to make other complex carbon compounds such as cellulose. Animals cannot make complex carbon compounds from simple inorganic ones like carbon dioxide, but rely on plants to supply them with organic compounds that are taken in when the plants are eaten. The carbon compounds made by plants and taken in by animals are said to be 'fixed'. They are returned to the environment by animals and plants in the carbon cycle.

The key stages in this cycle are:

- atmospheric carbon dioxide is fixed in plants during photosynthesis. The carbon dioxide is removed from the atmosphere.
- animals eat the plants and the carbon compounds are fixed in the animals.
- respiration in plant and animal cells returns the carbon compounds to the atmosphere. Plant material can also be burnt during combustion and turned into atmospheric carbon dioxide. Fossil fuels are the remains of plants and animals that died many millions of years ago. These fuels are burnt returning carbon dioxide to the atmosphere.

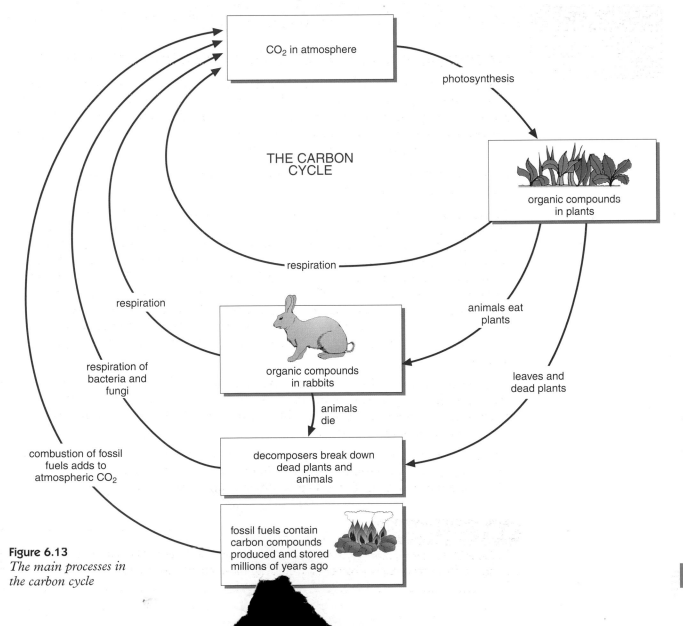

Figure 6.13
The main processes in the carbon cycle

Nitrogen cycle

Nitrogen is another element that must be recycled, as it is essential to the growth of animals and plants. Much of the nitrogen in plants and animals is in the form of amino acids and protein.

Did you know?

Neither plants nor animals can make direct use of atmospheric nitrogen gas. Some microbes can do this and these are called nitrogen-fixing bacteria. They are found in the soil where they turn nitrogen into nitrates which can be taken up by plant roots. The **nitrogen-fixing bacteria** may also be found in the root nodules of plants such as peas and beans. The microbes again turn the nitrogen into nitrates, which these plants can use.

Plants absorb nitrogen, in the form of nitrate ions, from the soil. Inside the plant, nitrates are combined with the organic compounds made during photosynthesis to form nitrogen-containing compounds such as proteins. When plants are eaten by animals some of the nitrogen in the plants become part of the proteins in their bodies.

Animals produce waste in the form of a nitrogen-containing compound called urea and this, together with dead plants and animals, is broken down by bacteria during a process called putrefaction. These putrefying bacteria produce ammonia. Plants cannot use ammonia as it is poisonous but the nitrogen becomes available to them once the nitrifying bacteria have converted the ammonium compounds into nitrates.

The main stages in the nitrogen cycle are summarised in Figure 6.14.

Figure 6.14
The main stages of the nitrogen cycle

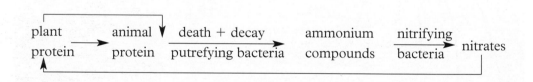

Summary

♦ Living things remove materials from the environment for growth and other life processes.

♦ Living things return the materials to the environment in their waste or when they die and **decay**.

♦ Materials decay by the action of micro-organisms.

♦ Decay is more rapid in moist, warm conditions where there is a plentiful supply of oxygen.

♦ Micro-organisms are used:
  – at sewage works to break down human waste
  – in compost heaps to break down plant material.

♦ The decay process releases substances which plants need to grow.

♦ In a stable community, the processes which use up materials are balanced by the processes which return materials. The materials are constantly being cycled.

♦ The constant cycling of carbon is called the carbon cycle.

♦ The constant cycling of nitrogen is called the nitrogen cycle.

Topic questions

1 a) What are detrivores? Name two.
b) What are decomposers? Name two.

2 Organic matter decays because of the action of some types of microorganisms.
a) Under what conditions do these microorganisms cause decay to be most rapid?
b) Where are microorganisms used to break down human waste?
c) Where are microorganisms used to break down plant waste?
d) Why is it important that organic matter decays?

3 What must be happening if an aquarium containing plants and animals is to remain clean and clear?

4 a) Which organisms remove carbon dioxide from the air?
b) What do green plants do with carbon dioxide?
c) Animals need carbon. From where do they get it?
d) How does carbon get released into the air?

5 a) In what form do plants absorb nitrogen from the soil?
b) What use do animals and plants make of nitrogen?
c) What do putrefying bacteria do?
d) What do nitrifying bacteria do?

Examination questions

1 The diagram shows some of the stages by which materials are cycled in living organisms.
a) In which of the stages, **A**, **B**, **C** or **D**:

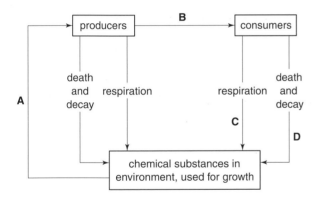

i) are substances broken down by micro-organisms.
ii) is carbon dioxide made into glucose.
iii) are plants eaten by animals? *(3 marks)*
b) In an experiment, samples of soil were put into four beakers. A dead leaf was put onto the soil in each beaker. The soil was kept in the conditions shown.

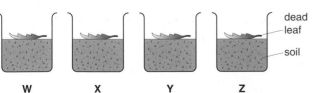

| **W** | **X** | **Y** | **Z** |
| warm and wet | cold and wet | cold and dry | warm and dry |

In which beaker, **W**, **X**, **Y** or **Z**, would the dead leaf decay quickest? *(1 mark)*

2 A population of rabbits lived on a small island. The graph shows their population over the last 50 years.

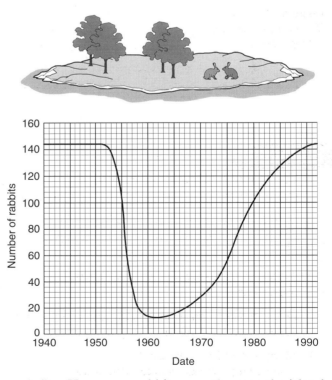

a) i) How many rabbits were there on the island in 1950? *(1 mark)*
ii) Give **one** year when there were 88 rabbits on the island. *(1 mark)*

125

b) i) Calculate the decrease in rabbit population between 1950 and 1960. *(1 mark)*
   ii) Suggest a reason why the rabbit population fell in these years. *(1 mark)*
c) The most rabbits on the island is always about 140. Suggest a reason for this. *(1 mark)*

3 Read the passage.

Glutton up a gum treet

Along the banks of the Cygnet River on Kangaroo Island, the branches of the dying gum trees stretch out like accusing fingers. They have no leaves. Birds search in vain for nectar-bearing flowers.

The scene, repeated mile upon mile, is an ecological nightmare. But, for once, the culprit is not human. Instead, it is one of the most appealing mammals on the planet – the koala. If the trees are to survive and provide a food source for the wildlife such as koalas that depend on them, more than 2000 koalas must die. If they are not removed the island's entire koala population will vanish.

Illegal killing has already started. Worried about soil erosion on the island, some farmers have gone for their guns. Why not catch 2000 koalas and take them to the mainland? "Almost impossible," says farmer Andrew Kelly. "Four rangers tried to catch some and in two days they got just six, and these fought, bit and scratched like fury."

---

a) Use the information from the passage and your own knowledge and understanding to give the arguments for and against killing koalas to reduce the koala population on Kangaroo Island. *(4 marks)*
b) The diagram shows the flow of energy through a koala.

The numbers show units of energy.

Respiration
12.25

Food
25.0

Growth
0.25

Faeces
12.5

i) Calculate the percentage of the food intake which is converted into new tissues for growth. Show your working. *(2 marks)*
ii) Give **three** different ways in which the koala use the energy released in respiration. *(3 marks)*

4 The information in the table compares two farms. Both are the same size, on similar land, close to one another and both are equally well managed.
a) Use this information to work out the average daily human energy requirement in kilojoules (kJ) per day. *(2 marks)*
b) The figures show that farms like Greenbank Farm can be nine times more efficient at meeting human food energy requirements than farms such as Oaktree Farm.
   i) The food chain for Greenbank Farm is:

   vegetation → humans

   What is the food chain for Oaktree Farm? *(1 mark)*
   ii) Explain why Greenbank Farm is much more efficient at meeting human food energy requirements. *(3 marks)*
c) The human population has been increasing rapidly throughout this century. It is now about 6 billion and is still growing. What does the information in this question suggest about likely changes in the human diet which may need to occur during the coming century? Explain your answer. *(4 marks)*

Name of farm	Activity	Energy value of food for humans produced in one one year	Number of people people whose energy requirements can be met by this food
Greenbank Farm	Grows food for humans	3285 million kJ	720
Oaktree Farm	Grows food for animals on the farm which become food for humans	365 million kJ	80

Chapter 7
Locomotion

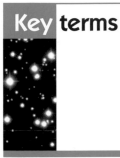

Key terms   aerobic respiration • aerofoil • antagonistic muscles • calcium phosphate • cartilage • compression • endoskeleton • exoskeleton • joints • lift • ligaments • locomotion • matrix • median fins • **muscle fibres** • **myotomes** • paired fins • pentadactyl limb • primary feathers • protein fibres • secondary feathers • skeletal tissue • sternum • swim bladder • **synovial fluid** • synovial joints • synovial membranes • tendons

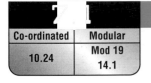

Co-ordinated	Modular
10.24	Mod 19 14.1

How vertebrates are adapted for movement

Locomotion

Locomotion is the ability of a living organism to move its body from place to place.

The importance of the skeleton

The bodies of all animals need support so that they are able to move from one place to another. For some soft-bodied animals, such as jellyfish, this support is provided by the water in which they live. For others, such as an earthworm or a caterpillar, the support is provided by the pressure of water inside the cells of their bodies and in the spaces between their body organs. The support for animals such as crabs and lobsters is provided by a hard shell, made up of a number of plates, on the outside of their bodies. This hard outside covering is called an **exoskeleton**.

Figure 7.1
A jellyfish in the water

Figure 7.2
A jellyfish out of the water

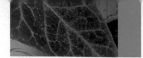

Figure 7.3
Caterpillars on a leaf

Figure 7.4
A crab shedding its shell

Vertebrates have an internal skeleton, called an **endoskeleton**. This type of skeleton grows as the animal grows.

This internal skeleton has three important functions:

● It provides rigid support for the soft parts of the body, so helping the body to keep its shape.

● It stops parts of the body from being damaged.

● It helps the organism to move from one place to another because the skeleton:
  – is made up of a large number of **joints** which allow the different parts of the body to move independently.
  – provides a firm anchorage for the muscles.

Did you know?

The human skeleton consists of about 200 separate bones and wherever bones meet, joints are formed. Some joints, such as those in the skull, do not allow bones to move. This is because the bones fit closely together.

The human skeleton has over 70 joints called **synovial joints**. They are given this name because the regions in the joint where bones might rub against one another are filled with a lubricating fluid called **synovial fluid**. Controlling the 70+ synovial joints are over 600 muscles.

There are three main types of synovial joints:

● *Sliding joints*, such as between the bones that make up the backbone. These allow movement in many directions

● *Ball and socket joints*, such as the hip and shoulder. These allow movement in many directions.

● *Hinge joints*, such as the knee, elbow, fingers and toes. These allow movement in one direction only.

Joints
What is in a synovial joint?

Synovial joints are adapted to provide as much friction-free movement between bones as possible.

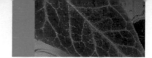

Figure 7.5, shows the important parts of a typical human synovial joint, in this case the shoulder joint.

Figure 7.5
The shoulder ball and socket joint

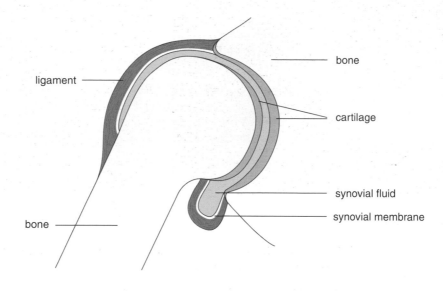

The adaptations of the features shown in Figure 7.5 for providing friction-free movement are contained in Figure 7.6.

Figure 7.6
The components of joints

Part	Function	Further information
Bones	Bones provide the solid base on which the tendons and ligaments are fixed.	Very resistant to being: ● **compressed**; ● bent; ● stretched.
Cartilage	**Cartilage** is the smooth layer which covers the ends of the bones and which stops the bones rubbing together.	● Is strong (high tensile strength), but not rigid. ● Can be compressed and is able to act as a shock absorber.
Ligaments	● **Ligaments** are strong fibres that hold bones firmly together. ● They form a protective cover around the joint.	Are very strong and sufficiently elastic to allow movement when the bones in the joint move, so reducing the chance of dislocating a joint.
Synovial fluid	● Synovial fluid is an oily liquid secreted by the synovial membrane. ● The fluid makes the surface of the cartilage slippery.	Helps the joint to move with as little friction as possible, by acting as a lubricant.
Synovial membranes	**Synovial membranes** secrete the synovial fluid.	Acts as a waterproof seal for the joint.
Tendons	**Tendons** attach muscles to bones.	Are very strong, but have very little elasticity and do not stretch, so that the muscle can pull on the bone.

More about bone

- Bone consists of large numbers of living bone-secreting cells in a matrix made up of protein fibres and calcium salts, mainly calcium phosphate. A matrix is solid material which cells secrete around them so that the cells are pushed apart, leading to cells being scattered in the matrix.

- If a bone is heated in a hot Bunsen flame for several minutes, the protein fibres and bone cells are destroyed. The bone is then very brittle and crumbles easily. The protein fibres and bone cells, therefore, stop bones from being brittle and help to strengthen them.

- Soaking a bone in acid for several hours will remove the calcium salts, leaving it soft and flexible. The calcium salts therefore provide the bone with its stiffness and hardness.

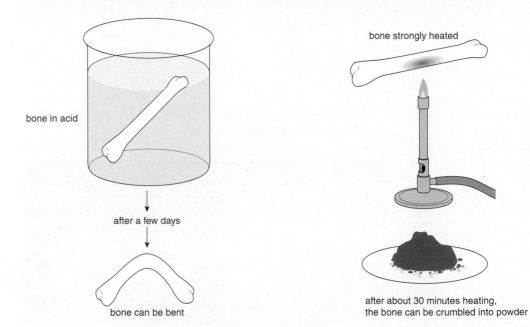

bone in acid

after a few days

bone can be bent

bone strongly heated

after about 30 minutes heating, the bone can be crumbled into powder

Figure 7.7
Experiments on bone

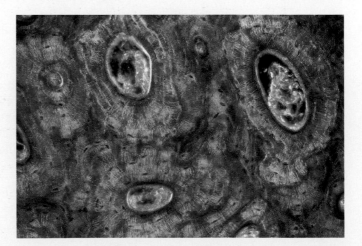

Figure 7.8
The bone matrix, as seen through a microscope

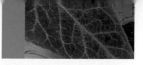

How exercise can benefit health

How muscles cause movement

Because muscles do work only when they contract, at each joint there has to be at least one muscle that contracts and moves the bone in one direction, and at least one other muscle which contracts and brings the bone back to its original position. In the elbow joint, the forearm is raised when the biceps muscle contracts and lowered when the triceps muscle contracts. Pairs of muscles, such as the biceps and the triceps, which work in opposite directions, are called **antagonistic muscle** pairs. When one muscle contracts, the other gets stretched.

The muscle tissue that makes bones move is called skeletal muscle. This muscle tissue is made up of a very large number of cylindrical **muscle fibres**.

Figure 7.9
The arrangement of muscle fibres in skeletal muscle

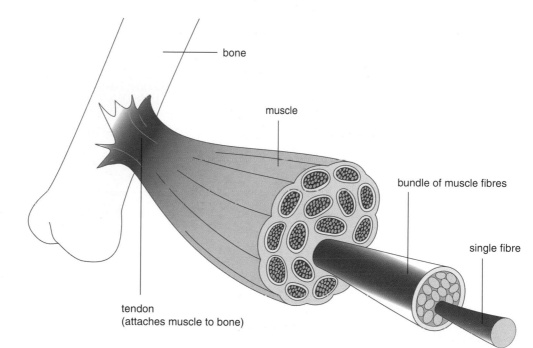

The muscle fibres can only do work when they contract. They contract only when stimulated by electrical impulses (see section 3.1). When the fibres contract, the muscle gets shorter and fatter. Although the muscle fibres can produce very powerful contractions, they tire quickly. This is why it is difficult to hold your arms above your head for a long time.

The energy needed to cause the contraction of the fibres is produced in the fibres during **aerobic respiration** (see section 2.5). Muscle fibres must have a good blood supply so that the glucose and oxygen needed for aerobic respiration (see section 2.5) can be transported to them; and so that the water, carbon dioxide and heat produced during respiration can be removed from them.

Why is regular exercise important?

In this country, it is estimated that about 40% of adults and 25% of people under 16 take almost no exercise.

Exercise contributes to good health and can improve the body in four ways:

1 *By improving muscle tone.* Muscles need to be slightly tensed so that they are ready to contract rapidly when required.

131

2 *By increasing muscle strength.* Regular exercise will improve a person's ability to carry out daily activities without muscles becoming stiff and sore.

3 *By increasing the flexibility of joints.* Joints and associated tissues that work efficiently are less likely to be damaged by such injuries as torn/pulled muscles, torn cartilages, sprains and dislocations.

4 *By increasing stamina.* An increase in stamina will provide for a more efficient heart, lung and circulatory system.

During any exercise, care has to be taken when using muscles that have not been used for some time. Movements that cause a sudden, large amount of stress on a particular joint, tendon, cartilage or bone often lead to one of several injuries.

Figure 7.10
Care must be taken to avoid over-stressing muscles and joints

Some injuries that affect the action of the joints and associated tissues

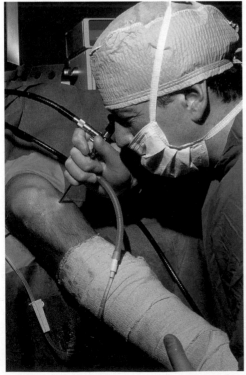

Figure 7.11
A surgeon using an arthroscope to examine a damaged knee joint

Figure 7.12
A damaged knee joint

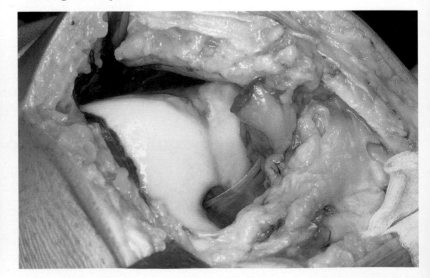

Figure 7.13
*Injuries associated with
damaged joints and
associated tissues*

Injury	What goes wrong	Other information
Dislocation	One of the bones in a joint is forced out of position.	Often caused by a fall, a blow, or extreme exertion.
Greenstick fracture	• The bone is not completely broken and there is no penetration of the skin. • Common in children, because their bones contain less calcium salts than the bones of adults.	Often caused by sudden impacts, extreme bending and twisting.
Simple fracture	• Bone completely broken. • The broken end(s) of the bone do not penetrate the skin.	
Compound fracture	• Bone completely broken. • The broken end(s) of the bone penetrate the skin.	
Shin splints	Inflammation (swelling) of the muscle that connects the top of the tibia with a bone of the foot.	Often follows lengthy, vigorous exercise.
Sprain	The twisting of a joint which results in the tearing of tendons and/or muscles.	Often caused by the sudden forcing of a joint into an unusual position, frequently caused by a sudden wrench.
Strain	The twisting of a joint which does not result in the tearing of tendons and/or muscles.	
Stress fracture	Hairline fractures in a bone.	In runners, it is caused by the compression of the bones in the leg due to the constant pounding of their feet on the ground.
Tendonitis	Tendons become inflamed, especially the Achilles tendon which joins the muscle connecting the tibia to a bone of the foot.	• Affects the knees in ball players, shoulders in swimmers and elbows in golfers and tennis players. • Caused by the muscles that are in constant use straining the tendons that hold the muscles to the bones.
Torn/pulled muscles	• The muscle fibres come apart. • Often occurs in the calf muscles and/or thigh muscles (hamstring).	Often caused by a sudden extra stress, such as violent kicking or hard running.

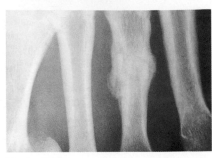

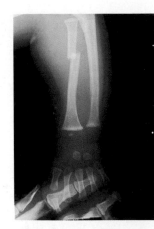

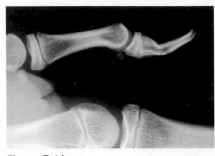

Figure 7.14
A stress fracture

Figure 7.16
A green-stick fracture

Figure 7.15
A simple fracture

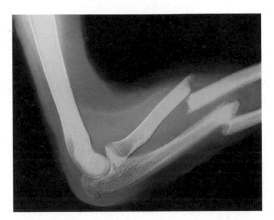

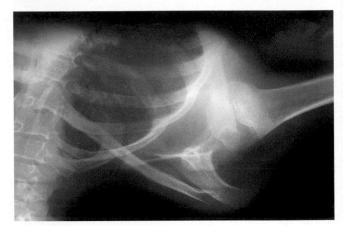

Figure 7.17
A compound fracture

Figure 7.18
A dislocated shoulder joint

Summary

- The internal skeleton of vertebrates provides the framework for support and movement.

- Bones are rigid to be able to support the body and provide a firm anchorage for muscles.

- **Joints** are formed wherever two or more bones come together.

- Joints that allow movement are called **synovial joints**.

- Each synovial joint contains **cartilage, ligaments,** bones, **synovial membrane** and **synovial fluid**. Each part is adapted to enable a joint to move easily.

- Muscles move bone by contracting.

- Muscle contraction requires energy from respiration.

- The harder muscles work, the faster the reactants and products of respiration need to be supplied and removed.

- Movement in unfamiliar situations can damage the tissues that make up a joint.

- Regular exercise keeps the body healthy in a variety of ways.

- Skeletal tissues, such as bone, cartilage, muscle, ligaments and tendons, have physical properties which adapt them for their functions.

- Bone is hardened by deposits of calcium phosphate.

- Living bone cells and protein fibres stop bone from being brittle.

- Ligaments have tensile strength and some elasticity.

- Cartilage is strong but not rigid.

- Tendons, which join muscles to bone, have tensile strength, but very little elasticity.

Topic questions

1 Give **three** functions of the human skeleton.

2 a) What is a synovial joint?
 b) What job does each of the following do in a joint
  i) tendon?
  ii) ligament?
  iii) synovial membrane?
  iv) synovial fluid?
  v) cartilage?

3 How is the structure of each of the following adapted to its function?
 a) Bone
 b) Cartilage
 c) Ligament
 d) Tendon

4 a) Describe the structure and action of skeletal muscle.
 b) Why do muscles need a good blood supply?

5 a) What happens when a bone is dislocated?
 b) What is the difference between a strain and a sprain?
 c) What happens when a muscle is torn?

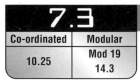

Co-ordinated	Modular
10.25	Mod 19 14.3

7.3 How fish are adapted for swimming

Adaptations of fish for swimming

Fish-like characteristics, such as body shape, a tail and muscles arranged in blocks, are all adaptations for swimming.

Body shape

Water offers very much more resistance to movement than does air. So any organism that moves through water has to transfer more energy. Most fish have a streamlined shape that allows them to move through the water with the minimum of effort.

Arrangement of muscles

Muscle tissue makes up a large proportion of the body weight of fish. The muscles are arranged in blocks on either side of the backbone. Figure 7.19 shows how these muscle blocks (**myotomes**) are arranged in a zig-zag pattern, one behind the other for the whole length of the body.

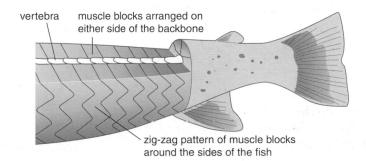

Figure 7.19
The arrangement of muscle blocks around the sides of a fish

Figure 7.20
The muscle blocks of a fish

The bones which make up the backbone of the fish are joined together to make a long flexible rod which ends at the tail. Contractions of the muscle blocks on one side of the body cause the backbone to bend because the muscle blocks immediately opposite, on the other side of the backbone, relax and stretch. The muscle blocks on opposite sides of the backbone act as antagonistic pairs of muscles.

The tail

The tail is a vertically arranged fin providing a large surface area. It is the main locomotory organ of most fish.

Muscles and swimming

During swimming, the muscle blocks on each side of the body contract and relax alternately and in a sequence that passes from the head to the tail. These movements of the muscles make the backbone bend, causing a series of wave-like movements to pass along the length of the fish from head to tail. The energy in these wave-like movements is transferred to the tail making it move rapidly from side to side. These side-to-side movements of the tail propel the fish forwards.

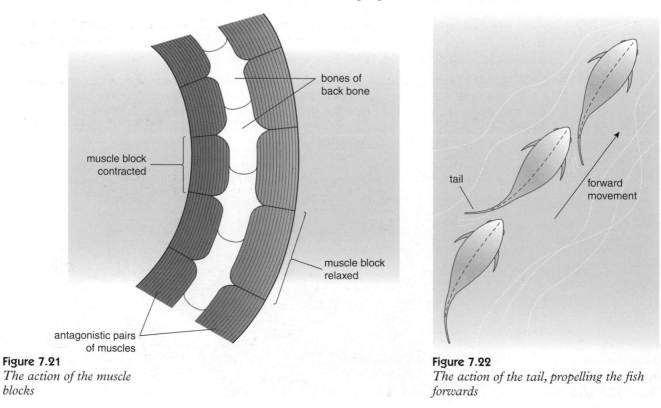

Figure 7.21
The action of the muscle blocks

Figure 7.22
The action of the tail, propelling the fish forwards

Figure 7.23
Top view of fish to show their shapes while swimming

More about swimming

Swimming is made more efficient by the actions of the fins and swim bladder.

The fins

Figure 7.24
The fins of a fish

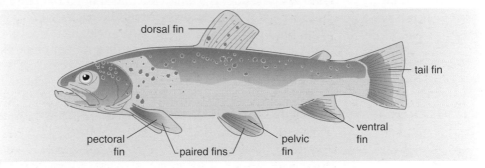

The fins allow a fish to control the position of its body.

- The median fins (the dorsal and ventral fins) help to keep the fish upright. These fins increase the surface area of the fish, thereby stopping the fish from rolling while it is moving through the water.

- By altering the horizontal angle of the paired fins (the pectoral and pelvic fins) in the water, fish can move upwards or downwards.

- By spreading the pectoral fins vertically and keeping them at right angles to the body, fish can use them as brakes.

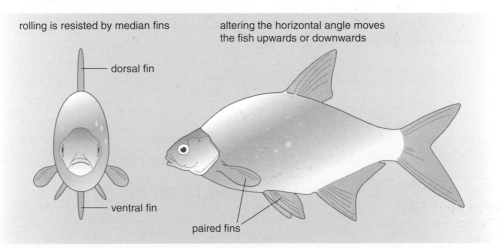

Figure 7.25
The action of the fins

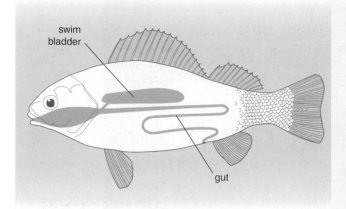

Figure 7.26
The position of the swim bladder

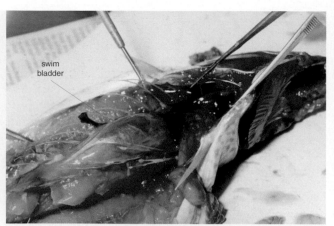

Figure 7.27
The swim bladder of a mackerel

The swim bladder

The swim bladder is a long, thin bag-shaped structure which is found inside the body cavity of fish, usually beneath the backbone. A fish can vary the amount of air in the swim bladder so that it can change its buoyancy as the pressure of the water varies with depth. Increasing the amount of air in the swim bladder increases the buoyancy of the fish (its body density decreases). Removing some of the air from the swim bladder decreases its buoyancy (its body density increases). By altering the amount of air in the swim bladder, the fish matches its density with that of the surrounding water. The fish is then weightless and can maintain its position, at different depths, using less energy. This makes it easier for the fish to catch its food, swim faster and be more agile.

Did you know?

- In some fish, such as herrings, the swim bladder is joined to the digestive system very near the mouth.

- In cod and trout, air passes into and out of the swim bladder through the walls of the blood vessels that surround it.

- In most fish the swim bladder can be inflated and deflated automatically, so that it can maintain buoyancy during sudden diving or surfacing.

- Sharks have no swim bladder so their bodies are always heavier than water. They start to sink as soon as they stop swimming.

7.4 How birds are adapted for flying

Co-ordinated	Modular
10.25	Mod 19 14.4

Adaptations of birds for flying

Although birds are adapted for life on land, their characteristic features, such as body shape, skeleton, wings and flight feathers, are adaptations for flight.

Body shape

In flight, birds tuck their feet into their bodies to give themselves as streamlined a shape as possible. Their feathers are arranged to overlap one another to provide a smooth surface from head to tail. This allows a smooth passage of air to flow over them, so reducing the friction between their bodies and the air.

Skeleton

If a bird is to remain airborne, then its body needs to be lifted by a force (called **lift**) which must be equal to or greater than its body weight. In order to reduce its body weight, the skeleton of a bird contains honeycombed bones which provide strength while reducing mass.

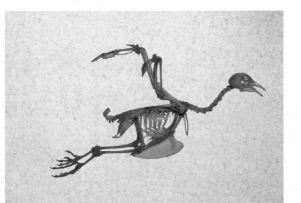

Figure 7.28
The skeleton of a pigeon

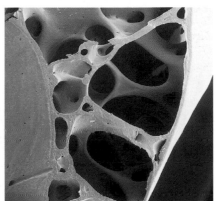

Figure 7.29
The honeycombed skull-bone of a bird

Wings

The up and down movements of the wings provide the force (lift) needed to allow a bird to fly. As the wings, with their large surface area, push downwards on the air, the bird is forced upwards and forwards.

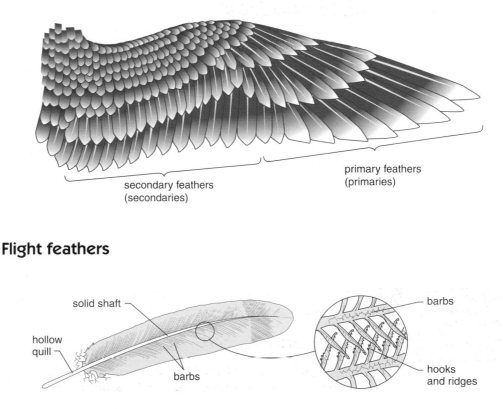

secondary feathers
(secondaries)

primary feathers
(primaries)

Figure 7.30
A bird's wing

Flight feathers

solid shaft

hollow
quill

barbs

barbs

hooks
and ridges

Figure 7.31
A bird's feather

139

Figure 7.32
Flight feathers

The flight feathers are the large feathers found on the wings. These feathers are very light and very strong.

More about flight

Flying is made more efficient by the shape of the wings, the structure and arrangement of flight feathers and the structure of the sternum (breastbone).

Shape of the wings

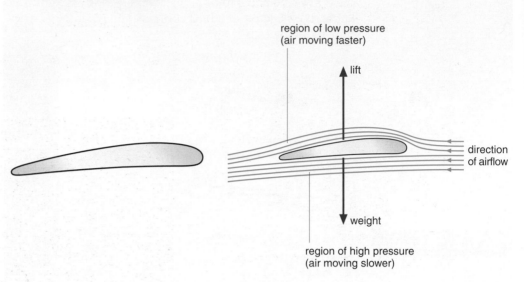

Figure 7.33
The aerofoil cross-section of a wing

Figure 7.34
Airflow over a wing

When viewed from the side, the wing of a bird is very slightly arch-shaped. This shape is called an aerofoil. Air flowing around such a shape produces an upward force on the aerofoil, causing it to lift. The upward force is created because air flowing across the top surface moves faster than air flowing under the wing. The faster the air flows, the lower the air pressure produced. So, the aerofoil shape of the wing produces lift because the air pressure above the wing is lower than that below the wing.

Structure and arrangement of the flight feathers

A flight feather consists of a large number of barbs, which are fixed on either side of a solid shaft. At the base of the feather, the shaft becomes a hollow quill. Together, these features produce a feather that is both light in weight, yet very strong. The barbs provide a flat surface to decrease air resistance during flight. To keep the barbs flat and together, there are tiny hooks on one side and tiny ridges on the other side (see Figure 7.31). The hooks fit into the ridges keeping the barbs in place. The hooks and ridges work like 'velcro'.

The secondary feathers (those attached to the ulna – a bone in the forearm) provide the aerofoil shape to the wings. The primary feathers (those fixed to the bones of the 'hand') can be moved independently from the secondary feathers by the action of the 'wrist' bones (see Figure 7.39).

Flying

In flight, the downbeat is the power stroke. During this stroke, the front (leading) edge of the wing is lower than the back (trailing) edge of the wing. The wing is providing not only lift, because the feathers are closed flat against each other, the wing is also pushing air backwards so making the bird go forwards.

wings moving down

On the downbeat the leading edge is lower than the trailing edge. Feathers are closed

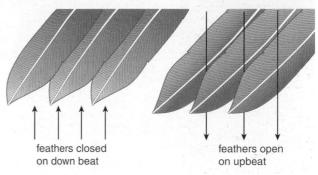

feathers closed on down beat

feathers open on upbeat

Figure 7.36
The arrangement of the feathers during flight

wings moving up

On the upbeat the primary feathers swivel open

Figure 7.35
The downbeat and upbeat of a wing

Figure 7.37
A bird in flight

On the upbeat, the primary feathers swivel open like the slats in a Venetian blind. This arrangement allows air to pass through the wing easily without reducing the lift produced during the downbeat.

The sternum

The muscles used to make the wings move are very large and very powerful. The wing muscles can make up as much as 1/5th of the total body weight of the bird. They need, therefore, to be attached firmly to a rigid and strong part of the skeleton.

Figure 7.38
The sternum of a bird

keeled sternum

The sternum is large relative to the size of the skeleton. This is because part of it has become adapted to form an enlarged, long ridge, called the keel, with a large surface area. It is on either side of the keel that each pair of flight muscles is fixed.

Pentadactyl limbs

The basic arrangement of the bones in the limbs of many terrestrial vertebrates is very similar. There is a general pattern in which similar bones are arranged in a similar order and position. This pattern is known as the pentadactyl limb (five-fingered limb). The fact that this general pattern exists in so many different vertebrates suggests that they have all evolved from a common ancestor. Any changes to this basic arrangement are adaptations to different environments and life styles.

Evolution of the wing

Wings are believed to have been adapted from the front legs of vertebrates that did not fly.

One obvious difference between the arm in humans and the wing in birds is that, in birds, the number of bones in the wrists and the hand are much reduced. This will have the effect of reducing the bird's total body weight.

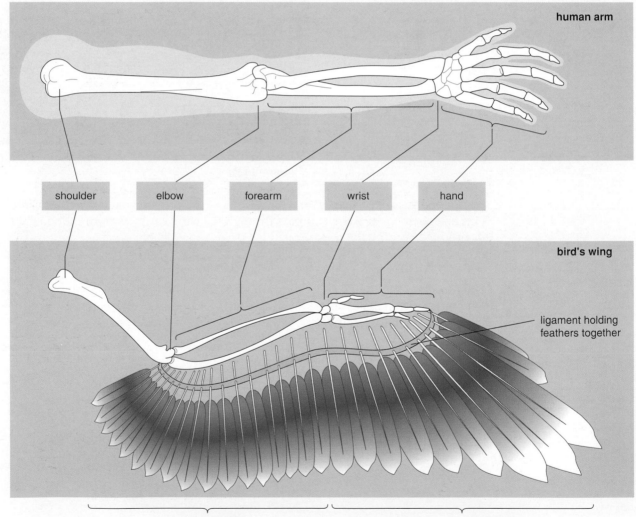

Figure 7.39
Pentadactyl limbs, a comparison of the bone arrangements in humans and birds

Did you know?

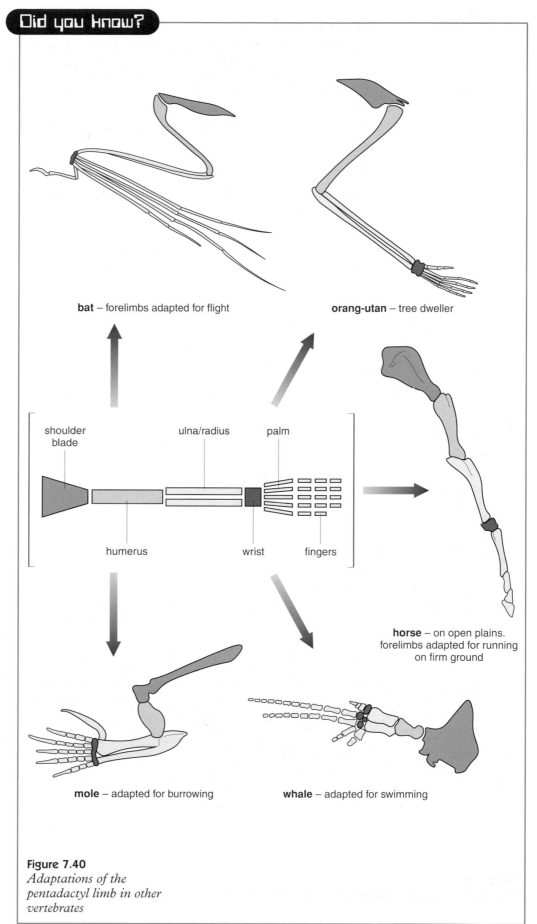

bat – forelimbs adapted for flight

orang-utan – tree dweller

shoulder blade

ulna/radius

palm

humerus

wrist

fingers

horse – on open plains. forelimbs adapted for running on firm ground

mole – adapted for burrowing

whale – adapted for swimming

Figure 7.40
Adaptations of the pentadactyl limb in other vertebrates

Summary

- The tail, body shape and arrangement of muscles are adaptations in fish for movement in water.

  - Swimming is assisted by the action of a swim bladder and the fins.

- Birds are adapted for flight by having wings, flight feathers, light bones and a streamlined shape.

  - Flying is assisted by the aerofoil shape of the wings, the design of the flight feathers and their arrangement during flying.

- The aerofoil wing shape provides the lift needed to keep the bird in the air.

- The flight feathers are adapted to provide a smooth surface.

- The flight feathers on the downbeat provide lift and forwards movement.

- The flight feathers on the upbeat open up to allow air to pass between them.

- The sternum has been adapted to provide a strong support for the wing muscles.

- The arrangement of bones in the wing is based on the pattern of the pentadactyl limb.

- The number of bones in the wrist and hands of birds is less than in a human limb.

Topic questions

1. a) How is the shape of a fish adapted to help it move through water?
   b) How is the shape of the tail adapted to help a fish move through water?
   c) How are the muscle blocks in the body of a fish arranged? How does the action of these muscle blocks help the fish to swim?

2. Which fins:
   a) stop the fish rolling?
   b) help the fish to move up or down?

3. a) What is the swim bladder? Where is it found?
   b) What is the function of the swim bladder? Explain how it carries out this function.

4. a) In which **two** ways is the skeleton of a bird adapted to help it fly?
   b) The shape of a bird's wing is described as being an aerofoil.
      i) What does this mean?
      ii) How does this shape provide the bird with lift as it flies?

5. Describe what happens to the wings and the wing feathers to help a bird fly during:
   a) the down beat.
   b) the upbeat.

6. Explain why many scientists think that birds and other vertebrates have evolved from a common ancestor.

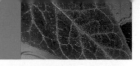

Examination questions

1 a) Movement is possible in humans because the skeleton has joints and muscles can pull on the bones at a joint.

The drawing shows the bones and some of the muscles in a human arm.

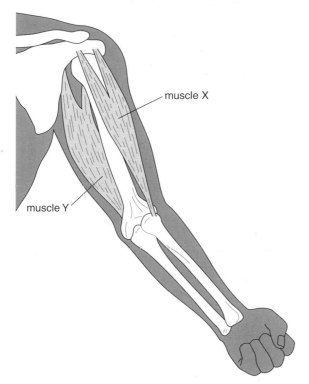

muscle X

muscle Y

i) Describe what will happen to the arm when muscle **X** contracts. *(1 mark)*

ii) Why is muscle **Y** needed? *(1 mark)*

iii) Choose words from the box to complete the sentences. You may use each word once or not at all.

elastic   hard   inelastic   rigid   tough

Ligaments hold joints together. To do this and allow movement, they must be _____ and _____ . Tendons attach muscles to bones. When muscles pull, the tendons transmit the pull to the bones. To do this, tendons must be _____ . *(3 marks)*

2 The drawing opposite shows some of the structures in a human knee joint.

a) Copy the diagram and label parts **A** and **B**. *(2 marks)*

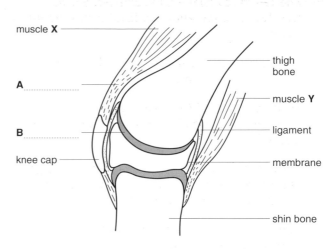

muscle **X**

thigh bone

A

muscle **Y**

B

ligament

knee cap

membrane

shin bone

b) Explain the function of the membrane in the joint. *(2 marks)*

c) A person suffers from a sprain in the knee joint. Describe what happens to the joint. *(2 marks)*

d) Explain how muscles **X** and **Y** bend and straighten the leg at the knee joint. *(3 marks)*

3 A student was given a chicken leg to examine.

a) The student observed that some ligaments were attached to the bone.

i) Give **one** property of the material which makes up ligaments. *(1 mark)*

ii) What is the function of the ligaments? *(2 marks)*

b) The student also recorded that tendons were attaching the muscles to the bone. When a tendon was pulled the claws moved. The tendon material must be strong. Give another property of the tendon material. Explain why this is necessary to move the bones at the joint. *(2 marks)*

4 a) List A gives the names of five parts of the arm. List B gives the jobs of these parts in a different order.

List A	**List B**
cartilage	makes the arm straight when it contracts
humerus	attaches a muscle to a bone
	covers the end of the humerus in the elbow joint
ligament	supports the upper arm
tendon	
	joins together the bones in the elbow joint
triceps	

Copy the lists. Draw a straight line from each part in List A to its job in List B. One has been done for you. *(4 marks)*

145

b) A runner falls and sprains his ankle joint. Describe how a joint is damaged when it is sprained. *(2 marks)*

c) Runners in a long race often have drinks containing the carbohydrate, glucose, during the race.

 i) Explain why runners need the carbohydrate, glucose, during a race. *(2 marks)*

 ii) It is better for a runner in a long race to have carbohydrates as glucose rather than as starch. Explain why. *(2 marks)*

5 The diagrams show two experiments which were done using chicken bones.

Experiment 1

bone

hydrochloric acid

after a few days bone removed and washed in water

bone can easily be bent

Experiment 2

bone held in clamp

bunsen flame

after twenty minutes bone removed and allowed to cool

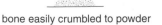
bone easily crumbled to powder

Two important substances in bone are protein fibres and calcium salts.

a) Explain how **Experiment 1** and **Experiment 2** show that these substances are in bone.
 Experiment 1:
 Experiment 2: *(2 marks)*

b) Explain why each of these substances is important in bone.
 Protein fibres:
 Calcium salts: *(2 marks)*

6 Fish swim using their tails to provide propulsion and their fins to stabilise themselves in the water. The speed at which they can swim depends upon a number of factors, including size, shape and the power they can generate.

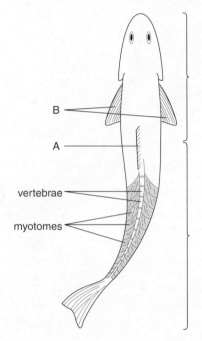

most internal organs in this region

B

A

vertebrae

myotomes

this region provides most propulsion

a) The diagram shows the position of the fins and the myotomes (blocks of muscle) in a typical bony fish.

 i) Name the fin labelled **A**. *(1 mark)*

 ii) State the function of the fins labelled **B**. *(1 mark)*

 iii) Suggest how the myotomes on opposite sides of the backbone behave as antagonistic pairs of muscles. *(2 marks)*

	Trout	Dace	Pike	Goldfish	Rudd	Barracuda
Length in m	0.24	0.16	0.17	0.10	0.22	1.20
Speed in m per s	2.60	1.60	1.70	1.10	1.30	12.00

b) The table gives information about the typical length and swimming speed of a number of fish.

   i) Plot the data on graph paper, with length in metres on the x axis and speed in metres per second on the y axis. Draw the most appropriate straight line through your plots. *(3 marks)*

   ii) The diagrams show the proportions of a trout and a rudd which are similar in length. Use the diagrams to explain the positions of the two fish on the graph. *(2 marks)*

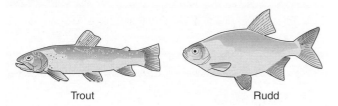

Trout   Rudd

7 In a bony fish, such as herring, the swim bladder is an air-filled sac lying above the gut. The volume of air in the swim bladder can be adjusted by swallowing air and this allows the fish to change its buoyancy as the pressure of water varies with depth.

a) Explain the advantages of the swim bladder to the herring. *(3 marks)*

b) Copy the diagram below, label with the letter **X** one fin which the herring uses to change from swimming close to the surface to swimming at a greater depth? *(1 mark)*

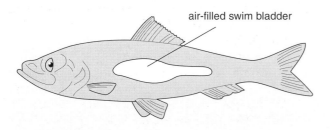

air-filled swim bladder

8 Birds and bats have skeletons which are both adapted for flight. The drawings show the body and wing structure of the small brown bat and the frigate bird. The drawings are not to the same scale.

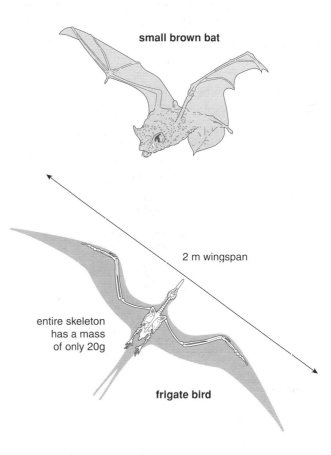

small brown bat

2 m wingspan

entire skeleton has a mass of only 20g

frigate bird

Use the drawings to help you to answer the questions.

a) How are the **wings** of the frigate bird adapted to give maximum lift? *(2 marks)*

b) The entire skeleton of the frigate bird has a mass of only 20 g. How is the skeleton modified to achieve this? *(1 mark)*

c) i) What do the wings of the small brown bat and the frigate bird have in common which adapts them for flight? *(1 mark)*

   ii) Describe **two** ways in which the wings of the small brown bat and the frigate bird are different. *(2 marks)*

Chapter 8

Feeding

8.1 How some invertebrates are adapted for feeding

Co-ordinated	Modular
10.26	Mod 19 14.5

Feeding adaptations in mussels

A mussel is an animal with a shell divided into two halves and hinged at one side. All mussels are aquatic, but most species live in seawater.

Mussels are usually found attached to rocks. They are described as being sedentary animals because during their lifetime they will not move very far. Because mussels stay in much the same area, their method of feeding has had to be adapted for the collecting of food from the water in which they live.

Mussels are **filter feeders**. These are animals that obtain their food by sieving the water and trapping any **plankton** it contains. Plankton are very small plants and animals that are floating in the water. The gills of the mussel are adapted not only for gaseous exchange, but they are also adapted for trapping plankton. At one end of the mussel's body are two siphons. These carry the water into and out of the closed shell. The gills have a large number of hair-like structures called cilia. These beat to draw in a current of water through one of the siphons.

Figure 8.1
Closed mussel shells

Figure 8.2
An open mussel shell

Figure 8.3
The feeding parts of a mussel

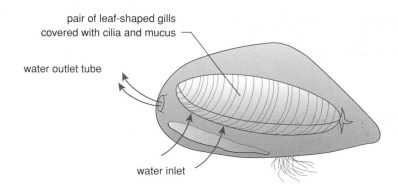

pair of leaf-shaped gills
covered with cilia and mucus

water outlet tube

water inlet

The gills sieve off plankton in this incoming water. The plankton become trapped in the sticky mucus produced on the surface of the gills. Some of the cilia on the mussel's gills are arranged to form a set of tracks. As these cilia beat, the food is moved towards the sorting palps and is eventually pushed into the mouth.

Did you know?

- One layer of mussels in a bath filled with seawater would filter the bath water more than 10 times in one hour.
- Many mussels are able to filter harmful bacteria from the water. So they can be used to control pollution. The collecting and eating of such mussels, however, could be fatal to humans.
- Even some whales are filter feeders.

Figure 8.4
A filter-feeding whale

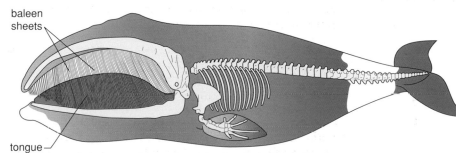

baleen
sheets

tongue

- These whales have thin sheets of a material called **baleen** – a strong flexible material made of the same protein as is found in human fingernails – hanging down from their upper jaw. The plates are arranged like teeth in a comb. During feeding the whale takes in huge amounts of water, then it closes its mouth and forces the water back out through the baleen plates. The plankton present in the water get trapped on the frayed inside edges of the baleen. The whale then sweeps its tongue along the back of the plates and swallows the plankton.

Feeding adaptations in mosquitoes

Mosquitoes are one of the many insects that are **fluid feeders**. Like all fluid feeders they need liquid food. Because mosquitoes live on a diet of blood, they need mouthparts that are adapted for:

- piercing the skin of the animal on whose blood they are feeding
- penetrating the blood capillaries
- sucking up the blood
- producing a chemical that will stop the blood from clotting.

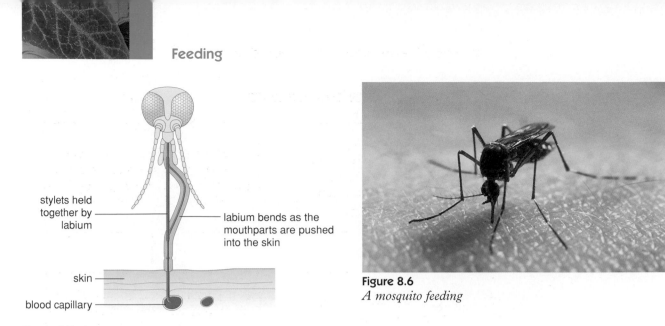

stylets held together by labium

labium bends as the mouthparts are pushed into the skin

skin

blood capillary

Figure 8.5
The mouthparts of a mosquito

Figure 8.6
A mosquito feeding

The mouthparts of the mosquito consist of:

- a number of needle-like tubes (**stylets**) for piercing the skin

- a needle-like stylet (food tube) for sucking up the blood

- a tube (the labium) with a very deep groove into which the other tubes fit when the insect is not feeding.

During feeding the needle-like stylets pierce the skin. The labium bends and acts as a guide for the needles. The food tube is so narrow that it could get quickly blocked if the blood being sucked up clots. To prevent this, saliva from the mosquito's salivary glands passes down the food tube into the blood in the capillary, before any blood is sucked up. The saliva contains a chemical called an **anticoagulant**. This stops the blood from clotting.

Malaria

Malaria is a disease of the blood. There are still many millions of cases of malaria occurring each year, about 1% of which are fatal. It occurs in many of the tropical and sub-tropical regions of the world, and is caused by an organism called *Plasmodium*. This organism is a single-celled parasite. A parasite is any organism that lives on or in another organism (the host) and obtains all its food from it. The malarial parasite has two hosts, the female mosquito called *Anopheles*, and humans. The female *Anopheles* mosquito carries the malarial parasite from host to host.

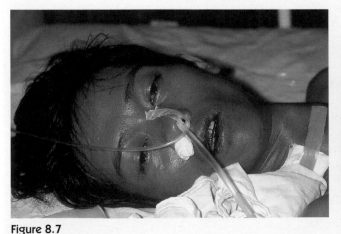

Figure 8.7
A malaria sufferer with a fever

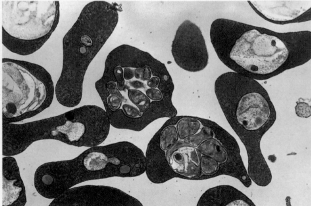

Figure 8.8
The malarial parasite in red blood cells

The life cycle of the malarial parasite

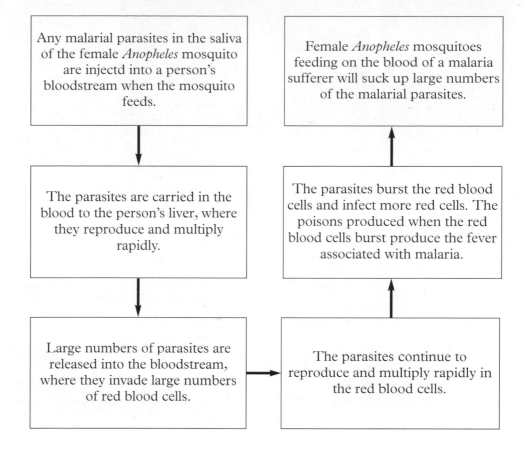

Any malarial parasites in the saliva of the female *Anopheles* mosquito are injectd into a person's bloodstream when the mosquito feeds.

Female *Anopheles* mosquitoes feeding on the blood of a malaria sufferer will suck up large numbers of the malarial parasites.

The parasites are carried in the blood to the person's liver, where they reproduce and multiply rapidly.

The parasites burst the red blood cells and infect more red cells. The poisons produced when the red blood cells burst produce the fever associated with malaria.

Large numbers of parasites are released into the bloodstream, where they invade large numbers of red blood cells.

The parasites continue to reproduce and multiply rapidly in the red blood cells.

Did you know?

- The insecticide DDT was developed in the 1930s mainly to kill mosquitoes to stop the spread of malaria. The insecticide is now known to be harmful to many other animals because it is non-biodegradable and its use in many countries is now prohibited.
- People who are carriers of sickle cell anaemia (see section 5.2) have some protection against malaria.

Feeding adaptations in other fluid feeders

Aphids

Aphids feed on the sap (a solution of sugars and amino acids) in the phloem vessels of plants (see section 4.3). Their mouthparts are adapted for piercing through plant material. Two sets of muscles are present. One set contracts to force the stylets into the plant tissue, the other contracts to pull the stylets out of the plant tissue. The needle-like stylets are sharp enough and long enough to penetrate the plant tissue and reach the phloem. The sap is sucked up along a groove between the sides of the two mouth parts.

these muscles pull the stylets out of plant tissue

these muscles force the stylets into plant tissue

muscles which guide stylets

stylets

Figure 8.9
Aphids

Figure 8.10
The mouthparts of an aphid

Butterflies

The mouthparts of butterflies are used to suck nectar (a dilute solution of sugar) from flowers.

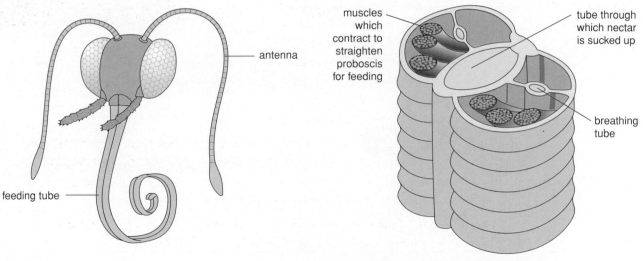

antenna

feeding tube

muscles which contract to straighten proboscis for feeding

tube through which nectar is sucked up

breathing tube

Figure 8.11
The mouthparts of a butterfly

Figure 8.12
A cross section of the butterfly feeding tube

Figure 8.13
A butterfly feeding

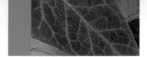

When not in use, the feeding tube is coiled tightly. Contractions of the muscles along its length uncoil the tube for feeding. The sugary solution of nectar is sucked up, with the feeding tube acting as a drinking straw.

Houseflies

Houseflies always seem to land on solid food, such as sugar, but they can only feed on food that is liquid. The mouthparts of a housefly are, therefore, adapted for dissolving the food before it can be eaten.

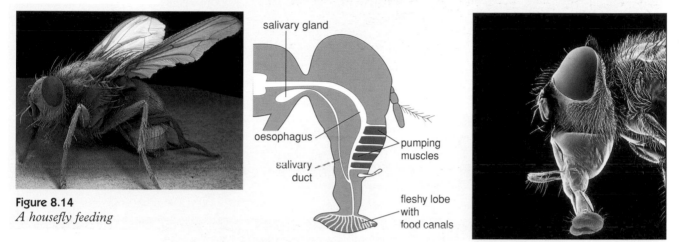

Figure 8.14
A housefly feeding

Figure 8.15
The mouthparts of a housefly

During feeding:

- saliva passes down the salivary duct.

- the saliva is sprayed onto the food through the canals in the fleshy lobe. The fleshy lobe has a large surface area, so saliva can act quickly on a large amount of food.

- the digestive juices in the saliva begin to digest the food.

- the action of the pumping muscles in the mouthparts, suck up the semi-digested liquid through the food canals in the fleshy lobe.

- if too much semi-digested liquid is sucked up, some may be forced out again as vomit.

Summary

◆ Mussels are **filter filters,** whose gill system has become adapted to filtering water to extract the **plankton** on which it feeds.

◆ Mosquitoes feed on blood. Their mouthparts have become adapted to piercing skin and sucking up blood.

◆ Malaria is a blood disease caused by a single-celled parasite that feeds and reproduces inside human red blood cells.

◆ Mosquitoes transmit malaria from one person to another as they feed.

◆ The mouthparts of an aphid, butterfly and housefly are adapted to suit their particular diets.

Topic questions

1 a) What are filter feeders?
  b) What are plankton?

2 a) How are the gills of a mussel adapted to help it trap food?
  b) How are the gills of a mussel adapted to get the trapped food into its mouth?

3 a) What are fluid feeders?
  b) Describe how the mouthparts in each of these insects help it to feed:
     i) mosquito

     ii) aphid
     iii) butterfly
     iv) housefly.

4 a) What is malaria? What causes it?
  b) What is a parasite?
  c) Describe the life cycle of the malarial parasite.

8.2

Co-ordinated	Modular
10.27	Mod 19
	14.6

How some mammals are adapted for feeding

Adaptation of teeth to their function in mammals

Mammals have teeth of various shapes and sizes. Some teeth are better adapted for biting off pieces of food, others are more suited for the tearing, gripping, chewing or crushing of food.

The types of teeth and the way they are arranged in the mouth is called the **dentition**. The dentition of most mammals has become adapted to suit their particular diets.

Dentition in humans

Human teeth have to cope with a wide variety of different foods from both animal and plant sources. Humans are described as being **omnivores**. Because of this, adult humans have four different types of teeth, each adapted for carrying out a different function.

incisor

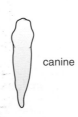
canine

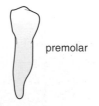
premolar

molar

Figure 8.16
Incisors are chisel shaped, and are used for biting off small pieces of food

Figure 8.17
Canines are long and pointed and used for tearing food

Figure 8.18
Premolars are large wide teeth, with a ridged surface, and are used to grind food into a moist paste

Figure 8.19
Molars are similar in shape and function to premolars

Figure 8.20
The human skull

Figure 8.21
*The arrangement of
teeth in the lower jaw*

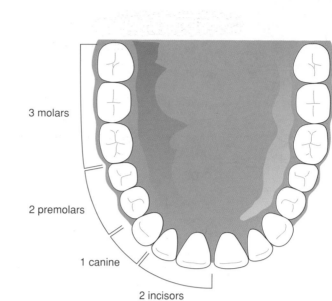

3 molars

2 premolars

1 canine

2 incisors

Figure 8.20 and Figure 8.21 show that teeth carrying out the same function are grouped together. This is to ensure that they can carry out their functions as efficiently as possible. The lower jaw in humans can move from side to side as well as up and down.

The numbers of each type of tooth in a jaw can be written as a **dental formula**. For an adult human with a full set of teeth, the dental formula would be written as:

$$I\frac{2}{2} \quad C\frac{1}{1} \quad P\frac{2}{2} \quad M\frac{3}{3} \quad = \quad 32$$

Dentition in dogs

Dogs are meat eaters – they are described as being **carnivores**. Their dentition, therefore, has become adapted to coping with their particular diet.

In a dog's jaw:

- the incisors are small, usually being used to cut off pieces of meat that are near the bone.

- the canines are large, sharp and pointed. They are used to grip and kill prey, and to pull the meat apart.

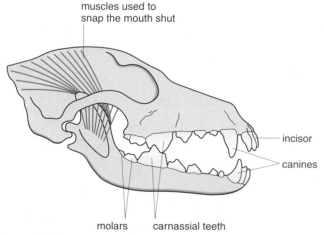

muscles used to
snap the mouth shut

incisor

canines

molars   carnassial teeth

Figure 8.22
The skull of a dog

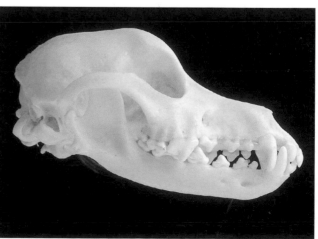

Figure 8.23
A dog feeding

- some of the premolars have become enlarged to form **carnassial teeth**. The top and bottom of these large teeth and the other premolars do not meet as they do in the jaw of a herbivore. In a dog and other carnivores, these teeth slide past each other like the blades on a pair of scissors. These teeth can be used for shearing meat from the bone and for cracking bones so that they can get at the bone marrow.

- the molars are relatively small as little time is spent in chewing the food.

The shape of the jaw bones only allows for an up and down movement of the lower jaw. It is this movement that provides the scissor action used when eating. The lower jaw is joined to the skull by very strong muscles. When these contract the mouth snaps shut.

The dental formula of a carnivore, such as a dog can be written as:

$$I\frac{3}{3} \quad C\frac{1}{1} \quad P\frac{4}{4} \quad M\frac{2}{3} \quad = \quad 42$$

Dentition of some other animals
Sheep

Sheep are **herbivores** because they have a diet consisting only of plant material. They spend much of the day feeding. This is because they need to digest a large quantity of plant material if they are to obtain the nutrients their diet requires. Their dentition has become adapted to allow them to graze.

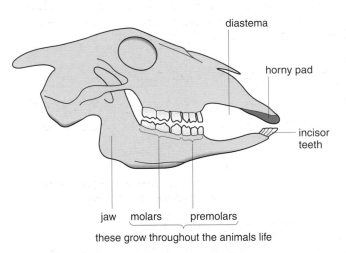

In the jaw of a herbivore, such as a sheep:

- the incisors in the bottom jaw are sharp and chisel-shaped.

- there are no incisors in the upper jaw. Instead, there is a hard, horny patch against which the lower incisors can bite.

- the canines are absent. Because their diet contains no meat these teeth are not needed.

- the space between the incisors and the premolars and molars is called the **diastema**. Sheep use their tongue to roll food around in this space.

- the premolars and molars are used for grinding. Because they are used so much, they wear down rapidly. They do not wear away though, because these teeth continue to grow for the whole of the animal's life.

The lower jaw can move up and down and from side to side.

The dental formula for a sheep can be written as:

$$I\frac{0}{3} \quad C\frac{0}{0} \quad P\frac{3}{3} \quad M\frac{3}{3} \quad = \quad 30$$

Figure 8.24
A sheep's skull

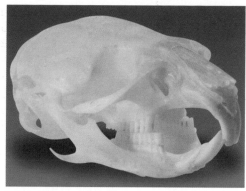

Figure 8.25
A walrus skull

Figure 8.26
A mouse skull

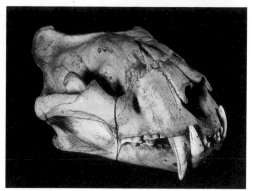

Figure 8.27
A tiger skull

The digestive system of some mammals

Just as the dentition of mammals has become adapted to their diet, so also have their digestive systems. Mammals that are herbivores eat large amounts of **cellulose**. However, the digestive system of mammals is not able to produce an **enzyme** that can break down cellulose into sugars. In order to break down the large amounts of cellulose that get eaten, parts of the digestive system of mammalian herbivores contain very large numbers of cellulose-digesting bacteria. These bacteria produce the enzyme (cellulase) that digests the cellulose into sugars. Some of these sugars are used by the bacteria themselves. The rest is used by the herbivore.

Digestion of cellulose in sheep and cows

The stomach in these herbivores is divided into four chambers. The 1st chamber (rumen) and 2nd chamber (reticulum) both contain very large numbers of the cellulose-digesting bacteria.

Did you know?

In many wild herbivores, such as deer, their feeding habits mean that they can eat quickly and then move to a safe place, from where they can chew the cud and keep watch for predators.

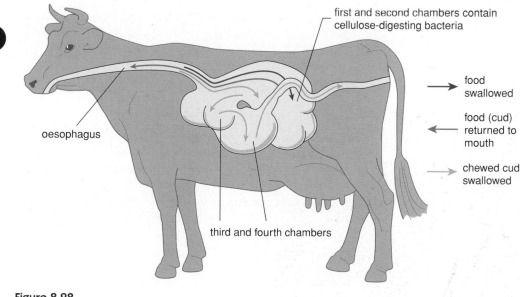

first and second chambers contain cellulose-digesting bacteria

oesophagus

third and fourth chambers

food swallowed

food (cud) returned to mouth

chewed cud swallowed

Figure 8.28
The digestive system of a cow

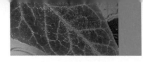

Figure 8.29
The four chambered stomach of a cow

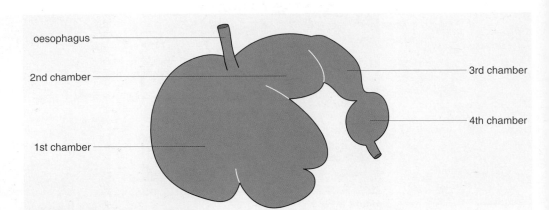

oesophagus

2nd chamber

3rd chamber

1st chamber

4th chamber

Figure 8.30
A cow chewing the cud

What happens during digestion?

- The grazed food is first chewed and mixed with saliva.

- It is then swallowed and passes into the 1st stomach chamber where it stays for several hours before passing into the 2nd chamber. In both these chambers are many millions of the bacteria which start the break down of cellulose.

- In the 2nd chamber the partially digested food is moulded into cud balls, and after some hours is regurgitated, as cud, back to the mouth.

- In the mouth the cud is chewed for up to 8 more hours ('chewing the cud').

- It is then re-swallowed, but this time passes into the 3rd chamber of the stomach. Here it is churned up and some water is extracted.

- It then passes into the 4th chamber, which is similar in its action to the stomach of humans.

- From here, it passes into the very long intestine. When it enters the intestine, digestion is still far from complete because the digestion of cellulose is a very slow process. This is why the intestine of a cow is about 50 m long.

Digestion of cellulose in rabbits

In rabbits the cellulose-digesting bacteria are found in a very large bag called the caecum, which opens into the digestive system at the junction of the small and large intestines (in humans, this bag is very small and seems to play no part in the process of digestion). The problem rabbits have to overcome is that by the time partially digested food leaves the caecum, it has already passed through the small intestine where the sugars are usually absorbed. In order to obtain the sugars needed for respiration, modifications to the generally accepted digestion process are needed.

Figure 8.31
The caecum and large intestine of a rabbit

small intestine

caecum

appendix

large intestine

Figure 8.32
Rabbits and their droppings

How does a rabbit get its sugars?

- Partially digested food leaves the caecum and enters the large intestine. The food has passed once through the stomach, small intestine and caecum, so the faeces are moist but still contain partially digested cellulose.

- During the night, a rabbit passes soft faeces from its anus. The rabbit eats these faeces and further digestion takes place.

- When the food reaches the small intestine for the second time, most of the products of the breakdown of cellulose can be absorbed into the bloodstream.

- In the large intestine most of the water is now removed from the faeces. These are passed out of the anus, as the dry pellets so often seen in areas inhabited by rabbits.

Carnivores and cellulose

Because the diet of carnivores does not include cellulose, they have no need for cellulose-digesting bacteria, nor do sections of the digestive system need to be adapted to contain them.

Mutualism or symbiosis

Mutualism (symbiosis) is the relationship between two organisms, usually of different species, in which both gain a benefit from living together. The relationship between the cellulose-digesting bacteria in the digestive system of herbivores is an example of this relationship. The herbivores benefit because they get the sugars they need. The bacteria benefit because they get a supply of cellulose and the nutrients they need.

Summary

◆ The shapes of the teeth of mammals are suited to their function.

◆ Human teeth are adapted to deal with a wide range of foods.

◆ Dogs have teeth that are adapted for a **carnivorous** diet.

◆ It is possible to relate the shape of teeth of most mammals to their function and particular diet.

◆ The digestive system of mammals is adapted to their diet.

◆ Mammals do not produce **enzymes** that can digest **cellulose**.

◆ Mammals with a **herbivorous** diet often have very large numbers of cellulose-digesting bacteria in their digestive systems.

◆ These bacteria digest cellulose into sugars.

◆ Sheep and cows' stomachs have chambers (one of which is the rumen) which contain these bacteria.

◆ To assist digestion, food is re-chewed in the mouth.

◆ Rabbits have cellulose-digesting bacteria in a large caecum. Because food digested in the caecum has already passed through the small intestine, rabbits eat their own faeces.

◆ Carnivores have no need for cellulose-digesting bacteria.

◆ The relationship between the cellulose-digesting bacteria and the herbivores is an example of mutualism (or symbiosis).

Topic questions

1 a) Why do humans have different types of teeth?
   b) What are the four types of human teeth? What shape is each? What is the main function of each?

2 What is the dental formula of an adult human who has a full set of teeth?

3 a) Describe the teeth of a dog, giving details of how they are adapted to a dog's diet.
   b) What is the dental formula of an adult dog?

4 a) Describe the teeth of a sheep, giving details of how they are adapted to a sheep's diet.
   b) What is the dental formula of an adult sheep?

5 a) How is the digestive system of a herbivore adapted to its diet?
   b) Describe how a cow digests its food.
   c) Why do rabbits eat their own faeces?

6 The relationship between bacteria in the stomach of a cow and the cow itself is an example of mutualism. Explain why.

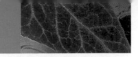

Examination questions

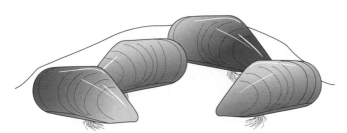

1 A mussel is a mollusc found firmly attached to rocks at the sea shore, as shown in the diagram.

Mussel opened to show the gills on one side.

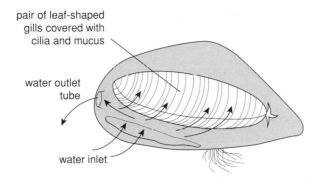

pair of leaf-shaped gills covered with cilia and mucus

water outlet tube

water inlet

Explain how the mussel feeds. *(3 marks)*

2 The aphid (or greenfly) extracts sugary sap from the phloem of young plants tissues. The diagrams show an aphid and an enlargement of the mouth parts.

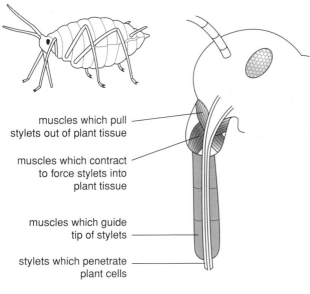

muscles which pull stylets out of plant tissue

muscles which contract to force stylets into plant tissue

muscles which guide tip of stylets

stylets which penetrate plant cells

Give **two** ways in which the mouth parts are adapted for this function. *(2 marks)*

3 Cows are large animals that feed naturally by grazing in fields of grass. After grazing they can be seen lying down "chewing cud".
a) Copy and complete the following statements. Digestion starts in the mouth. The cow has ———— teeth for cropping or pulling the grass. The cow grinds the grass using its ———— teeth. *(2 marks)*
b) The digaram shows the stomach region of a cow. The stomach of the cow is in four parts and the order in which the food passes through is explained in the boxes.

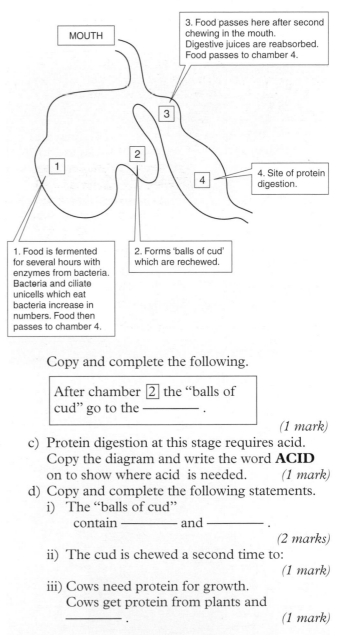

MOUTH

3. Food passes here after second chewing in the mouth. Digestive juices are reabsorbed. Food passes to chamber 4.

3

2

1

4

4. Site of protein digestion.

1. Food is fermented for several hours with enzymes from bacteria. Bacteria and ciliate unicells which eat bacteria increase in numbers. Food then passes to chamber 4.

2. Forms 'balls of cud' which are rechewed.

Copy and complete the following.

After chamber ② the "balls of cud" go to the ———— .

(1 mark)
c) Protein digestion at this stage requires acid. Copy the diagram and write the word **ACID** on to show where acid is needed. *(1 mark)*
d) Copy and complete the following statements.
   i) The "balls of cud" contain ———— and ———— . *(2 marks)*
   ii) The cud is chewed a second time to: *(1 mark)*
   iii) Cows need protein for growth. Cows get protein from plants and ———— . *(1 mark)*

4 Some insects have mouthparts which are specialised to feed on fluids.
Draw **one** line from each box to touch the structure that it describes. Box 1 has been done for you.

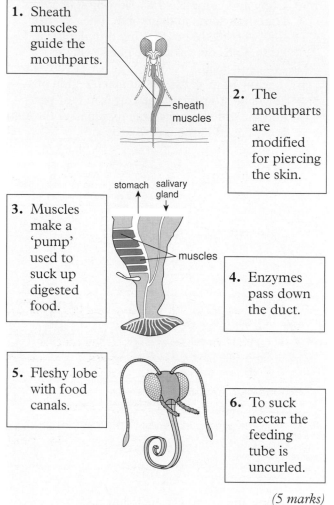

1. Sheath muscles guide the mouthparts.

sheath muscles

2. The mouthparts are modified for piercing the skin.

stomach salivary gland

3. Muscles make a 'pump' used to suck up digested food.

muscles

4. Enzymes pass down the duct.

5. Fleshy lobe with food canals.

6. To suck nectar the feeding tube is uncurled.

(5 marks)

5 Most mammals feed using teeth.
a) The drawings show the arrangement of the teeth in the lower and uper jaws of an adult human.

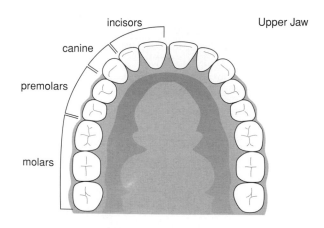

incisors   Upper Jaw
canine
premolars
molars

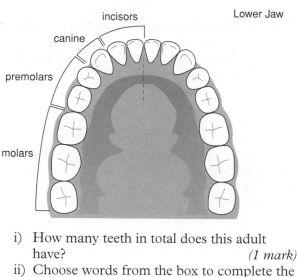

incisors   Lower Jaw
canine
premolars
molars

i) How many teeth in total does this adult have? *(1 mark)*
ii) Choose words from the box to complete the sentences. You may use each word once or not at all.

| bite | chew | front |
| pointed | ridged | side |

The incisor teeth are placed at the ———— of the mouth and their function is to ———— off pieces from the food. Human canine teeth are smaller in proportion and less ———— than those of a carnivore. Their molar teeth are less ———— than those of a herbivore. *(4 marks)*

b) The dental formula for an adult human with all their teeth would be written as:

$$i\frac{2}{2} \quad c\frac{1}{1} \quad p\frac{2}{2} \quad m\frac{3}{3}$$

The back four molars are sometimes called wisdom teeth. Write the dental formula for someone who has had all their wisdom teeth taken out.

i– c– p– m– *(1 mark)*

6 Tooth shape and jaw actions are suited to the different diets of mammals.

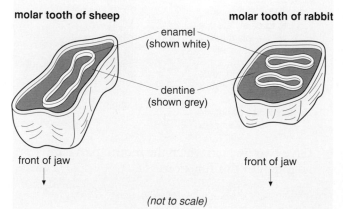

molar tooth of sheep   **molar tooth of rabbit**

enamel (shown white)

dentine (shown grey)

front of jaw   front of jaw

(not to scale)

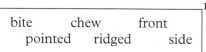

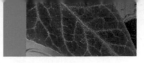

a) Look at the diagrams of the molar teeth of the sheep and rabbit.

 i) Examine the direction of the ridges. In which direction do you think that the rabbit moves its jaw as it uses its molar teeth?

(1 mark)

 ii) Why do herbivores have broad molar teeth with ridges of enamel? *(1 mark)*

b) The following diagrams show the position of the teeth in the upper and lower jaws of the rabbit.

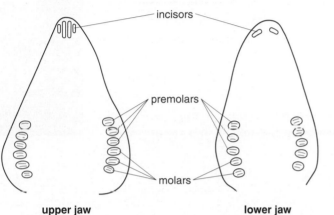

upper jaw   **lower jaw**

Examine the upper and lower jaws of the rabbit and complete the dental formula for the rabbit.

  i– c– pm– m–

(1 mark)

c) The digestive system of the rabbit is also adapted to its diet.

rabbit stomach and intestines

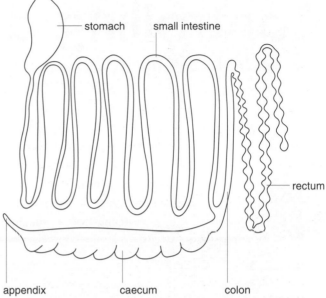

The food passes into the caecum that lies between the small intestine and the colon.

 i) Explain how the bacteria in the caecum help the rabbit digest the grass. *(2 marks)*

At night a rabbit produces soft faeces which it immediately eats.

 ii) Suggest why eating the night faeces is an important part of rabbit nutrition.

(2 marks)

Chapter 9
Controlling disease

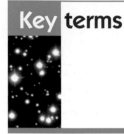

9.1

Co-ordinated	Modular
10.28	Mod 20 15.1

How biology has helped us to control infectious disease

How Louis Pasteur showed that decay and disease are caused by living organisms

Up until the 1800s it was thought that whatever made food go bad was made by the food itself. For example, people believed that maggots were created by the action of meat rotting. This idea was called spontaneous generation. Even after the discovery of a wide range of bacteria, it was still thought that these organisms were created by the rotting substances in which they were found, causing food to decay.

A French scientist, Louis Pasteur (1822–1895), could not accept the idea of spontaneous generation. In 1854, Pasteur carried out some work on the fermentation of yeast and showed that the bacteria causing wine to go sour could be killed if the sugar solutions were heated before the yeast was added. This led him to believe that when food went bad it was not caused by the bacteria in the food itself, but by the bacteria in the air that reached the food.

Figure 9.1
Louis Pasteur

In 1861 Pasteur began a series of investigations to prove his theory. For the food he used a clear nutrient broth. When left exposed to the air the broth would go bad in a very short time. He knew he could heat the broth to kill any bacteria in it, but he had to devise a way of allowing the broth to be exposed to the air but not to any of the bacteria that were in the air. To solve this, he designed special glass flasks each with a long thin neck that could later be heated and bent into an S-shape.

Figure 9.2
Pasteur's experiment

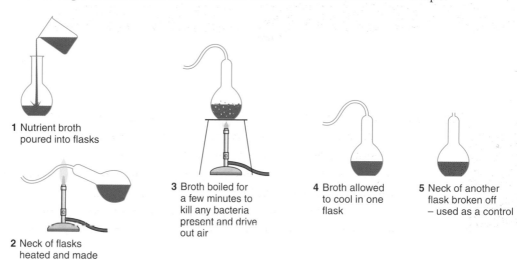

1 Nutrient broth poured into flasks

2 Neck of flasks heated and made into S shape

3 Broth boiled for a few minutes to kill any bacteria present and drive out air

4 Broth allowed to cool in one flask

5 Neck of another flask broken off – used as a control

Figure 9.3
One of Pasteur's original flasks

In carrying out his investigations he:

- poured the broth into a glass flask;
- heated the neck and bent it into an S-shape;
- boiled the broth to kill any bacteria in it and get rid of most of the air.

He believed that as the flask cooled down, air would be drawn back into the flask, but that any bacteria in the air would be trapped in the bend in the neck of the flask. Bacteria were, he considered, heavier than air.

In order to make his investigations as scientific as possible, he:

- used a control experiment in which, after boiling the broth, he cut off the long thin neck of the flask to allow air to reach the broth;

- carried out a large number of repeats.

The results supported Pasteur's original theory. None of the broth in the flasks with the S-shaped necks went bad. The broth in the flasks with the open tops went bad within a few days.

Pasteur spent most of the rest of his life working on the causes of various diseases, in particular, cholera and rabies, and their prevention by means of vaccination.

> ### Did you know?
>
> - Pasteur's discovery led to the pasteurisation of milk. In this process, milk is heated to about 72°C for 15 seconds and then cooled rapidly. This treatment kills any bacteria in the milk but does not spoil its taste.
>
> - Some years after Pasteur's work on broth, a German doctor, Robert Koch (1848–1910) discovered the causes of many more diseases. He took blood samples from people who had a particular disease and grew the microorganisms in the blood sample on a special nutrient jelly, called agar. He then injected the microorganisms into mice and discovered that they got the same disease.

The treatment of diseases: the use of antibiotics

By the end of the 19th century, due to the work of Pasteur and Koch, the particular microorganisms causing many diseases were known. Microorganisms that cause disease are called **pathogens**. Most, but not all, pathogens are either **bacteria** or **viruses** (see section 3.5).

Many diseases can be treated with medicines, such as **penicillin** (see section 10.2) and other **antibiotics**. These contain chemicals (drugs) that kill the disease-causing bacteria that get into the body.

Some medicines contain chemicals, such as aspirin, paracetamol or codeine. These relieve the symptoms of an illness, such as a high temperature or aches and pains. However they do not cure the disease because the drugs have no effect on the pathogens themselves.

Figure 9.4
Some commonly used medicines

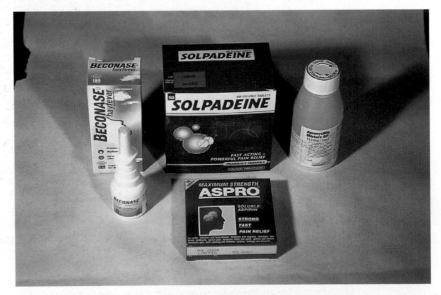

For many years antibiotics have been a successful way of treating many bacterial infections. However, they do not work against disease-causing viruses (**viral pathogens**).

Figure 9.5
Bacterial and viral diseases

Diseases caused by bacteria	Diseases caused by viruses
salmonella (food-poisoning)	common cold
tuberculosis (TB)	influenza (flu)
gonorrhoea	rabies
pneumonia	measles
typhoid fever	mumps
dysentery	rubella
syphilis	AIDS
diphtheria	herpes (e.g. cold sores)

Viruses can only live and reproduce inside living cells. Because of this, antibiotics cannot get to a virus to have any effect on it. Many drugs that are developed to kill viral pathogens are unlikely to work unless they also destroy the body's cells in which the viruses are living and reproducing. However, there are a few anti-viral drugs that can treat viral infections in the early stages. For example, Zovirax, which is used to treat cold sores in the early stages.

Did you know?

In 1928, Alexander Fleming, a Scottish bacteriologist working at St Mary's Hospital in London, was growing colonies of bacteria on plates of agar (see section 10.3). Some of these plates were accidentally left uncovered for several days. When he examined these plates, Fleming noticed that there were colonies of a fungus growing on the agar. To his surprise, he also noticed that where the fungus was growing, the colonies of bacteria had been killed. The fungus was a mould called *Penicillium*. Fleming believed that the fungus was producing a chemical that could kill bacteria.

Figure 9.6
Fleming's plates showing penicillin and bacteria

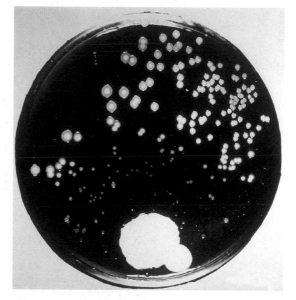

It was not until about 12 years later that the chemical was isolated and given the name 'penicillin'. Penicillin was, therefore, the first antibiotic. It began to be used in hospitals in 1940. The majority of patients who would normally have died from bacterial infections recovered as a result.

The penicillin age

- When first used, penicillin was considered to be the 'wonder drug'.

- By the 1950s strains of bacteria were beginning to appear that were unaffected by penicillin. This was not considered to be much of a problem because new antibiotics were being developed to cope with the resistant strains.

- During the next 20 years, more and more bacteria were discovered to be resistant to penicillin. Although the drug companies continued to develop some new generations of antibiotics, many of the companies had turned their attention to the treatment of viral infections.

Did you know?

- The bacterium *Streptococcus pneumonia* causes diseases such as bacterial pneumonia, meningitis, middle ear infections and sinusitis.

Figure 9.7
Streptococcus
pneumonia
bacterium

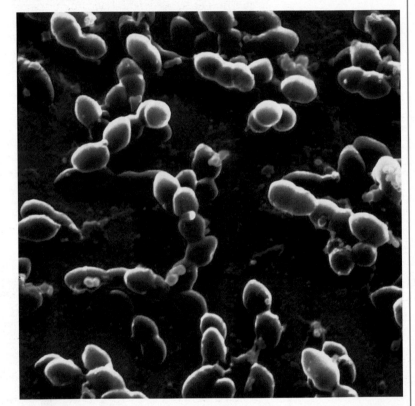

Until the late 1970s penicillin was effective in killing this bacterium. It is now estimated that at least 30% of infections caused by this bacterium are penicillin-resistant. Other diseases, such as tuberculosis, gonorrhoea, syphilis and salmonella, are also becoming resistant.

- The bacterium *Staphylococcus aureus*, is found on the skin of every person. It is normally harmless but can cause infections in wounds. Because this bacterium is now becoming resistant to most antibiotics, including a powerful antibiotic called methicillin, it is known as MRSA (methicillin-resistant *Staphylococcus aureus*). MRSA causes problems in hospitals to patients who are elderly, very sick or who have open wounds. This is because if the bacterium gets into the bloodstream it can cause a number of serious infections. The treatment of MRSA involves a high standard of hygiene, including the careful washing of the hands of the patient and hospital staff, isolation of the patient and treatment using expensive antibiotics, such as vancomycin. In 1997, a *Staphylococcus aureus* bacterium partially resistant to even this antibiotic was discovered.

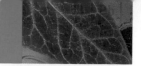

How do bacteria become resistant to antibiotics?

It is important to realise that it is the bacteria that become resistant to antibiotics not the patient!

Figure 9.8
Bacterium with the long strands of its DNA clearly visible

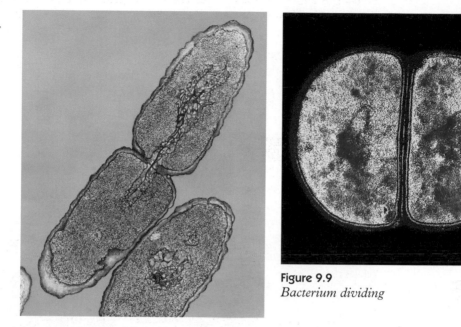

Figure 9.9
Bacterium dividing

Bacteria reproduce by dividing. This can take between several minutes or several hours, depending on the particular bacterium and the environmental conditions. Such rapid rates of cell division frequently give rise to **mutations** of the bacterial DNA (see sections 5.1 and 5.3).

Many antibiotics work by deactivating (switching off) a particular bacterial protein (a target protein). The genetic changes caused by a mutation of the bacterial DNA can result in:

● the loss of the target protein so the antibiotic has no target to deactivate;

● changes to the target protein which prevent the antibiotic from binding to the bacterium;

● the bacterium producing so many target proteins that the antibiotic cannot deactivate them all;

● the bacterium producing an antibiotic-deactivating enzyme.

When a patient takes an antibiotic, every single bacterium in their body is exposed to its effects. The antibiotic will kill or stop the growth of most of the bacteria. However, if a few bacteria contain mutated bacterial DNA that makes them resistant to the effects of the antibiotic, then these bacteria will grow and reproduce. In a short time, they become the dominant strain in that patient's body. Further treatment with the same antibiotic will not be effective against this new strain of bacteria.

Misuse of antibiotics

Unnecessary prescriptions

Many patients still expect their doctor to prescribe antibiotics for colds and flu, which are caused by viruses; and chest infections, some of which are also caused by viruses. The antibiotic will not cure the illness because it has no effect on viruses.

Instead the prescribing of an antibiotic for an infection caused by a virus is likely to lead to the production in the patient's body of antibiotic resistant strains of bacteria. If a bacterial infection is suspected, then the doctor may collect a sample of the bacteria from the infected site and have the bacteria identified. Once identified, the most effective antibiotic for that strain of bacteria can then be prescribed.

Not finishing a course of treatment

Many patients stop taking the antibiotics when they begin to feel better. Often in such cases not all the bacteria causing the infection will have yet been destroyed. This can give rise to a resistant strain of bacteria that may not respond to the same antibiotic if the infection returns. You should always finish a course of antibiotics.

Figure 9.10
A nurse taking a throat swab

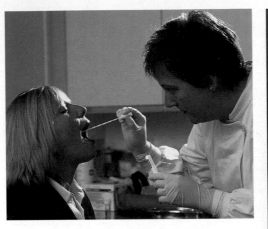

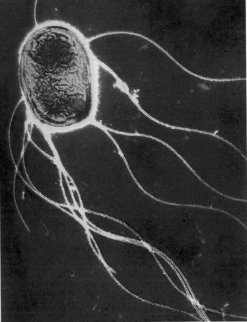

Figure 9.11
Salmonella *bacterium*

Use of antibiotics in the production of food

Since the 1960s antibiotics have been used as growth promoters in the intensive rearing of poultry, pigs and cattle. The users believe that the antibiotics help the animals get fatter quicker and eat less food, although there is a lack of scientific evidence to support these beliefs. This practice has led to the rise of antibiotic-resistant strains of bacteria that cause food poisoning in humans. Because the antibiotics fed to the animals are similar to those used to treat food poisoning infections in humans, many cases of food poisoning are difficult to treat.

Although new antibiotics are being developed, there are fewer available today than there were 30 years ago, so it is important that we use these antibiotics sensibly.

The treatment of diseases: immunisation

Figure 9.12
Instructions for a course of antibiotic treatment

If bacterial or viral pathogens get into the bloodstream, they are detected by the white blood cells (see section 2.2). On the outside of these pathogens are proteins called **antigens**. These antigens cause certain white cells to produce proteins called **antibodies**. These antibodies attach themselves to the antigens and, hopefully, destroy the disease-causing pathogens. Our bodies can make over a million different antibodies. This is because there are a very large number of different antigens. The antibodies can provide a **natural immunity** against repeated infections.

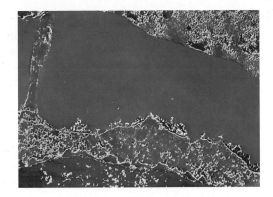

Figure 9.13
Antigens (uneven red areas) on the surface of a white blood cell (bottom)

The main problem involved with acquiring a natural immunity against a particular disease is that you first have to have the disease. This can cause serious risks to health if the disease is a dangerous one, such as smallpox. The problem is overcome by being immunised.

When you are immunised against a particular disease, the doctor will introduce small quantities of inactive or dead forms of the particular pathogen into your bloodstream. These inactive or dead pathogens are called the **vaccine** and the process is called **immunisation**, also now known as **vaccination**. Immunisation can be done by injection, as with measles, mumps, rubella and tetanus; by a scratch made in the skin, as with smallpox; or by mouth, as with polio.

As soon as the vaccine gets into the bloodstream, the antigens, which will still be on the outside of the inactive or dead pathogens, stimulate the white cells to produce the required type of antibodies. The vaccine will have been prepared carefully to make sure that the patient gets either a mild form of the particular disease or, as with most immunisations, no disease at all.

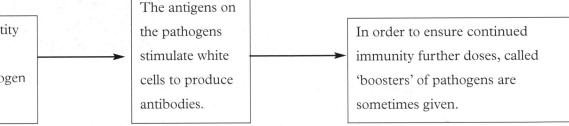

A small quantity of dead or inactive pathogen is given.	→	The antigens on the pathogens stimulate white cells to produce antibodies.	→	In order to ensure continued immunity further doses, called 'boosters' of pathogens are sometimes given.

Without immunisation it takes your body about one week to produce a sufficient number of antibodies to counteract the effects of a new antigen. With immunisation your body can respond by rapidly making the correct antibody, in the same way as if you had already had the disease. This response is called **active immunity**.

MMR vaccination

A combined measles–mumps–rubella (MMR) vaccine was first used in 1968. Each of the three diseases is caused by a viral pathogen. The vaccine, which is given by injection, contains live virus particles of the three pathogens, that have been modified to stop them giving the person the full effects of the diseases. Two doses of MMR vaccine are now given to UK children, the first when they are 12–15 months old and the second before they start school.

Emergency treatment

- Rabies is a viral infection of the nervous system. It is nearly always fatal if not treated. The rabies pathogen usually gets into the bloodstream through the bite of an infected animal.

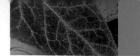

In such a situation, the person who has been bitten by an infected animal requires immediate treatment. This involves the person being injected with rabies anti-serum which contains ready-made antibodies. This is called **passive immunity**. Because the person has not produced the antibodies in the serum, the body will gradually remove these from the bloodstream. Passive immunity does not, therefore, provide long term protection.

Figure 9.14
A rabies infected raccoon

● Tetanus is a bacterial infection of the nervous system. The tetanus bacteria are commonly found in soil, dust and manure. Cuts such as those made by rusty nails, dirty knives or gravel are particularly susceptible to becoming infected with tetanus bacteria. If a person is diagnosed as having the disease then doses of tetanus anti-serum are given. These contain the anti-tetanus antibodies. Long term protection (active immunity) against tetanus can be provided by vaccination.

Did you know?

● Rabies is also known as hydrophobia ('*hydro*' is the Greek word for 'water', '*phobos*' is the Greek word for 'fear'). It was given this name because although the infection makes patients very thirsty, they find it very difficult to drink water because the muscles in their food pipe go into spasms even at the sight of water.

● Tetanus, also known as lockjaw, causes severe muscle spasms and tightening of the chest, neck and jaw muscles. The tightening of the neck and jaw muscles can be so extreme that the patient is unable to open his or her mouth to swallow food or water.

● Both rabies and tetanus anti-serums are produced by immunising horses with the appropriate dead or inactive pathogen, and then purifying and concentrating the horse's plasma which will contain the desired antibodies.

How the treatment of disease has changed as a result of increased understanding of the action of antibodies and immunity

The various ways in which the treatment of a disease such as tuberculosis (TB) has changed, shows clearly the links between an increase in scientific knowledge, and understanding of the causes and methods of prevention of diseases.

What is tuberculosis?

Tuberculosis (TB) is an infectious disease that is spread by droplet infection (coughing and sneezing). It is caused by a bacterial pathogen. TB can affect several organs of the body, including the brain, kidneys and bones, however most commonly it affects the lungs. If untreated, the disease destroys the tissues of the lungs. At present about 3 million people die from TB each year.

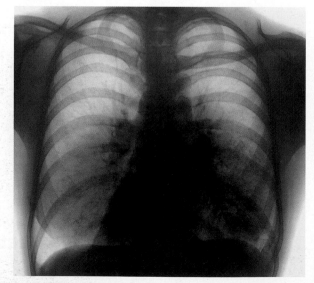

Figure 9.15
An x-ray of the lungs of a person suffering from tuberculosis

Figure 9.16
Scientific landmarks in the treatment of tuberculosis

Date	Scientific landmarks
9000–5000BC	Skeletal remains from the Neolithic era show evidence of TB.
3700–1000BC	Skeletal remains from Egyptian mummies show that TB was widespread.
20	Celsus, a Roman physician, prescribed sea voyages as an early treatment for TB.
850–1000	Arabian physicians prescribed a mixture of camphor, sugar distilled from grapes, and dry air as a treatement for TB.
1720	An English physician was the first to consider that TB was caused by 'minute living creatures'.
1854–1950	Introduction of hospitals in the countryside where TB patients could get plenty of fresh air. These hospitals were called sanatoria (singular: sanatorium).
1865	A French military surgeon showed that TB was due to a specific microorganism and that it was passed between humans and cattle.
1882	Robert Koch discovered the bacteria responsible for TB.
1895	William Roentgen discovered X-rays. These enabled doctors to follow the progress of the disease in patients.
1924	French bacteriologists Calmette and Guerin, developed the vaccine know as BCG. This vaccine is still used to treat TB today.
1944	The antibiotic streptomycin was discovered which was more effective in treating TB than penicillin was.
Present day	TB bacteria are becoming resistant to most of the drugs in use. Rather than use one drug at a time, the latest treatments involve several drugs being used at once. This is done in the belief that the bacteria cannot produce mutations that resist the combined action of several drugs used together.

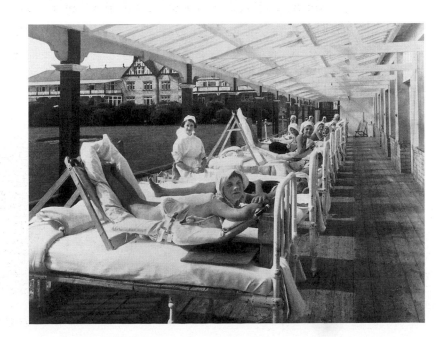

Figure 9.17
TB patients taking in the fresh air at a sanatorium

The advantages and disadvantages of being vaccinated against a particular disease

As a responsible parent, having to decide whether to have your child immunised against many of the infections which they are likely to come into contact with, requires a knowledge and understanding of the advantages and disadvantages of the vaccinations. The advantages and disadvantages can be illustrated by considering the MMR vaccinations.

Measles is a fever producing disease that is very infectious, especially in children. Complications of the disease can include:

- inflammation of the brain (meningitis);
- inflammation of the brain (encephalitis);
- bronchitis;
- pneumonia;
- convulsions (fits).

Figure 9.18
Complications associated with measles

Complications	Possible risk from the infection
pneumonia/bronchitis	1 in 25
convulsions (fits)	1 in 200
meningitis/encephalitis	1 in 1000
death	1 in 2500 to 5000

Mumps is an infectious disease which causes glands in the neck to swell. Complications of mumps include:

- inflammation of the testicles of young males; if it occurs after puberty this can lead to infertility;
- deafness;
- inflammation of the brain (encephalitis);
- inflammation of the central nervous system (meningitis).

Figure 9.19
Complications associated with mumps

Complications	Possible risk from the infection
swollen, painful testicles	1 in 5
deafness (usually gets partly or completely better)	1 in 25
meningitis/encephalitis	1 in 200 to 5000

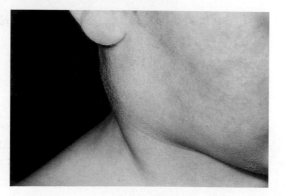

Figure 9.20
The swollen glands of a child suffering from mumps

Rubella, also called German measles, is a mild disease. However, if a pregnant woman catches rubella, especially in the first three months of the pregnancy, the virus can cause serious damage to the nervous system of the developing fetus.

Figure 9.21
Complications associated with rubella

Complications	Possible risk from the infection
damage to unborn fetus	• 9 out of 10 pregnancies (in the first 8 to 10 weeks) • 1 in 5 to 10 (between 10 and 16 weeks) • After 16 weeks damage is rare

Figure 9.22
The rash caused by rubella

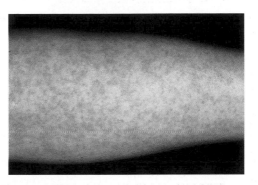

The MMR vaccine, like all vaccines, may not be 100% safe. However, as the contents of the above tables show, the diseases themselves can produce some very unwelcome complications. The decrease in risk from some of these complications if the vaccine is given, is shown in the table below.

Figure 9.23
A comparison of the risks of complications from the infection and from the vaccine

Complications	Possible risk from the infection	Possible risk from the vaccine
convulsions	1 in 200	1 in 1000
meningitis/encephalitis	1 in 200 to 1 in 5000	1 in 1 000 000
death	1 in 8000 to 1 in 10000 (depends on age)	0

(The contents of the tables are based on information provided by Medinfo.co.uk)

It is important to realise that without vaccination not all children are able to cope with having the actual diseases. Vaccination provides such children with the chance of being protected against any unwanted complications. In the case of the MMR vaccine it also provides protection for unborn babies from rubella.

How white blood cells provide immunity

Blood contains two different types of white cells (lymphocytes) which are involved in providing immunity against infection. These are known as T cells and B cells.

T cells

The surface membrane of each T cell contains thousands of identical T cell receptors. If the body is invaded by a particular pathogen, the antigens on the outside of the pathogen are recognised by the receptors. The T cells bind to the antigens on the invading cells surface and destroy the cell. Each T cell is specific for a particular antigen.

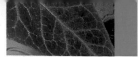

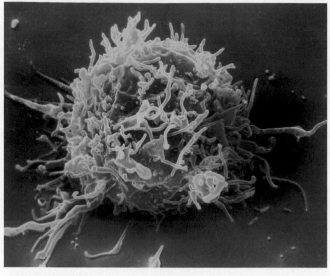

Figure 9.24
T cell

Figure 9.25
B cell

B cells

B cells produce antibodies which are Y-shaped proteins. Each and every one of the millions of B cells in our blood has a different antibody on its surface. Each antibody has a specific structure that will recognise a particular antigen.

Figure 9.26
Y shaped antibodies

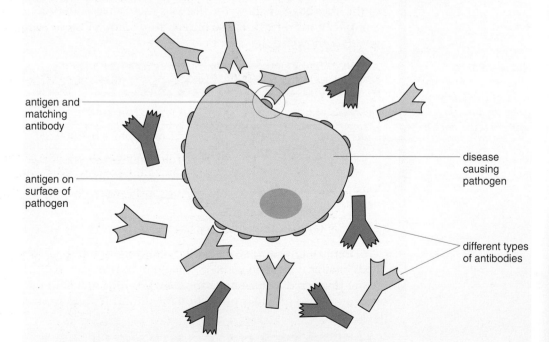

antigen and matching antibody

antigen on surface of pathogen

disease causing pathogen

different types of antibodies

T cells can also stimulate B cells to multiply rapidly to form clones (see section 5.1). Each of the clones can then produce more antibodies.

After an infection certain B cells, called **memory cells**, remain in the body. These can then trigger a rapid production of appropriate antibodies to even low levels of a particular antigen if the same infection reoccurs. It is this immunological memory that provides the rapid response to repeated infections, so giving immunity following a natural infection or immunisation.

Summary

◆ Microorganisms that cause diseases are called **pathogens**.

◆ Some chemicals, called **antibiotics**, kill bacterial pathogens.

◆ Antibiotics will not kill viral pathogens.

◆ Some drugs treat the symptoms and not the infection.

◆ Bacteria can develop resistance to antibiotics.

◆ **Immunisation** (**vaccination**) introduces small quantities of dead or inactive forms of the pathogen (**vaccine**) into the body.

◆ The vaccines contain **antigens** which cause the white cells to produce **antibodies**.

◆ The body can be made to be **actively** or **passively immune** to infection.

◆ The MMR vaccine protects against measles, mumps and rubella.

◆ T cells are lymphocytes which contain receptors that destroy antigens. These cells can also stimulate the production of antibodies.

◆ B cells are lymphocytes that produce antibodies.

◆ Antibodies are specific to a particular antigen.

◆ After an infection memory cells remain in the body so that antibodies can be made rapidly if the same infection reoccurs.

Topic questions

1 a) What did Pasteur's experiments on nutrient broth show about what caused food to go bad?
   b) Give two ways in which Pasteur carried out his work scientifically.

2 a) What are pathogens?
   b) What are antibiotics?
   c) Against which type of pathogens do antibiotics have no effect?

3 Why, if you got flu:
   a) should you not be prescribed antibiotics?
   b) might you take aspirins?

4 a) Why, when first used, was penicillin considered to be a wonder drug?
   b) Why is penicillin no longer thought to be a wonder drug?

5 Describe how bacteria can become resistant to the effects of antibiotics.

6 a) What are antigens? Where are they found?
   b) What are antibodies? What do they do?

7 During immunisation a vaccine is introduced into your body.
   a) What does the vaccine contain?
   b) How does the immunisation protect you?

8 a) What is the difference between active immunity and passive immunity?
   b) When are you likely to be given passive immunity?

9 a) What are lymphocytes?
   b) What are T cells? How do these fight infection?
   c) What are B cells? How do they fight infection?
   d) What are memory cells? How do they fight infection?

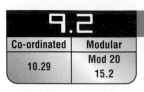

9.2	
Co-ordinated	Modular
10.29	Mod 20 15.2

How biology has helped us to treat kidney disease

Details of the structure and function of the kidneys are found in section 3.4.

What happens if the kidneys fail to work properly?

The failure of one kidney causes few problems. However, if both kidneys fail there is no control over the composition of the blood, and the amount of urea present in it will increase to such a level that the person becomes poisoned by his or her own waste products. If this situation is left untreated death can occur quite quickly.

Kidney failure can be treated in two ways:

● The patient can be connected to a **dialysis machine**, which acts like an artificial kidney.

● The patient can be given a **kidney transplant**. This is an operation in which the person's damaged kidney is replaced with a healthy kidney from someone else (called a **donor**). The person receiving the healthy kidney is called the **recipient**.

There are two types of kidney transplants:

– living donor transplants, in which a living person is willing to donate a kidney to someone with kidney failure;

– cadaveric transplants, in which the kidney comes from a person who has just died.

The dialysis machine

Details of the action of a dialysis machine are given in section 3.4.

During dialysis it is important that useful substances, for example, glucose and ions of minerals such as sodium, are not lost from the blood as it passes through the machine. In order to stop this happening the dialysis solution is made up to match the concentrations of these substances found in the patient's blood plasma. In this way only waste substances and excess ions and water diffuse into the dialysis solution across a **partially permeable membrane**.

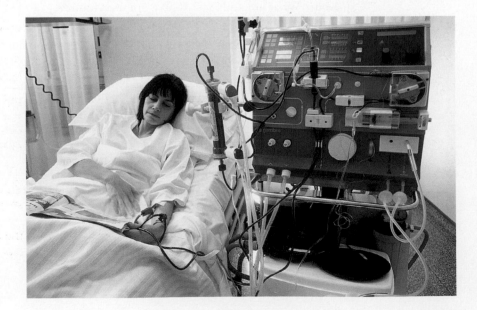

Figure 9.27
A person connected to a dialysis machine

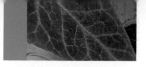

Kidney transplants

Some details of kidney transplants are given in section 3.4.

The main problem to be overcome in this method of treating kidney failure is that the recipient's immune system will treat the new kidney as a foreign body and try to destroy it. This happens because the recipient's T cells do not recognise the antigens on the cells of the donated kidney as 'self'. In order to reduce the risk of rejection of the healthy kidney, a number of things need to be done:

Figure 9.28
A kidney ready for transplant

- the recipient can be given drugs to suppress their immune system. This makes the body less able to fight infections;

- a donor with a 'tissue-type' (genetic makeup) similar to that of the recipient should be found. The ideal donor would be an identical twin, because both would be genetically identical;

- the bone marrow of the patient can be treated with radiation. This would stop the production of white blood cells, so suppressing the immune system;

- the recipient can be kept in sterile areas to avoid the risk of any infections.

The advantages and disadvantages of treating kidney failure by dialysis or kidney transplants

The choice of treatment for kidney failure depends upon what is available, and what is appropriate. Each treatment has its own requirements, advantages, and disadvantages that need to be considered carefully before any choice can be made.

Figure 9.29
The advantages and disadvantages of kidney transplants and dialysis

Treatment	Advantages	Disadvantages
Transplant	<ul><li>no dialysis needed.</li><li>normal diet.</li><li>fewer hospital visits after the treatment.</li><li>more normal life-style.</li></ul>	<ul><li>stress involved in waiting for a kidney match.</li><li>no suitable match may be found.</li><li>risk associated with surgery.</li><li>daily medication needed to prevent rejection.</li><li>lower resistance to infections because of immuno-suppressant drugs.</li></ul>
• Living donor transplant	<ul><li>90–95% transplant survival rate one year after surgery.</li><li>operation can be scheduled to suit recipient and donor.</li></ul>	<ul><li>donor has to undergo surgery.</li><li>donor must be perfectly healthy.</li></ul>
• Cadaveric transplants	88% transplant survival rate one year after surgery.	<ul><li>patient needs to be on a waiting list until a suitable kidney becomes available.</li><li>timing of the operation is dependent on the availability of a suitable kidney.</li></ul>
Dialysis	<ul><li>no rejection problems.</li><li>more readily available than kidneys are for transplants.</li><li>no risks from surgery.</li></ul>	<ul><li>very controlled diet.</li><li>may need to be attached to the machine for 10 hours a day 3 times a week.</li></ul>

During a kidney transplant, as with most operations, a blood transfusion may be required. The blood being transfused must match the blood group of the recipient.

Blood groups

There are four main groups to which a person's blood can belong. These are known as A, B, AB and O. The letters refer to the particular antigen that is present on the surface of the red blood cells.

- Group **A** blood has type **A** antigens on the surface of its red cells.
- Group **B** blood has type **B** antigens.
- Group **AB** blood has type **A** antigens and type **B** antigens.
- Group **O** blood has **no** antigens.

The plasma contains antibodies that will attack the antigen not present on the red blood cells.

- Group **A** blood has **anti-B** antibodies in the plasma.
- Group **B** blood has **anti-A** antibodies.
- Group **AB** blood has **neither** anti-A nor anti-B antibodies.
- Group **O** blood has both anti-A and anti-B antibodies.

If the same type of antigen and antibody were present they would combine and cause the red blood cells to clump (agglutination). This clumping can lead to blocked blood vessels and even death.

Figure 9.30

Agglutination following a blood transfusion

Blood group A
Type A antigens on the donor's red blood cells

Blood group B
Anti-A antibodies in the recipient's plasma

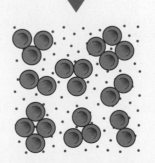

mixed together during transfusion

recipient's antibodies cause the donor's red blood cells to clump together

Figure 9.31
Summary of blood groups

Blood group	Antigen on red blood cell	Antibody in plasma
A	A	anti-B
B	B	anti-A
AB	A and B	neither
O	neither	anti-A and anti-B

To ensure clumping does not take place during a blood transfusion, it is important to make sure that the patient's plasma does not contain the antibody against an antigen carried on the donor's red blood cells. The table below shows which blood transfusions can be carried out safely.

Figure 9.32
Blood group compatibility

Patient's blood group	Donor's blood group			
	A (anti-B)	B (anti-A)	AB	O (anti-A + anti-B)
A (anti-B)	✓	✗	✗	✓
B (anti-A)	✗	✓	✗	✓
AB	✓	✓	✓	✓
O (anti-A 1 anti-B)	✗	✗	✗	✓

✓ = no clumping   ✗ = clumping

When checking to find out if clumping will occur, only the donor's antigens and the patient's antibodies need to be considered.

Summary

◆ Kidney failure can be treated either by using a **dialysis machine** or by having a **kidney transplant**.

◆ In a dialysis machine blood flows between **partially permeable membranes**.

◆ The concentrations of glucose and mineral ions in a dialysis machine match those of the patient's blood plasma.

◆ Various precautions need to be taken to avoid the rejection of a transplanted kidney.

◆ During a blood transfusion there must be compatibility between the blood of the donor and that of the patient.

◆ This compatibility is based on the ABO system of blood grouping.

Topic questions

1 a) Why do people become ill if both their kidneys fail to work?
   b) Kidney transplants are only given to patients when both kidneys fail to work. Why?

2 a) Care has be taken after a kidney transplant that it is not rejected. Why might it be rejected?
   b) Describe four ways to reduce the possibility of kidney rejection. For each, state how the chances of rejection are reduced.
   c) Why would rejection be unlikely if the donor and recipient were identical twins?

3 How, in a kidney dialysis machine, is glucose prevented from being lost from the patient's blood?

4 In a blood transfusion it is important that clotting does not occur.
   a) What name is given to this effect? Why is it dangerous?
   b) What causes the blood to clot during a blood transfusion?
   c) Complete the gaps in the following table to show the antigens and antibodies associated with each of the four blood groups. Some parts have been done for you.

Blood group	Antigen on red blood cell	Antibody in plasma
A	A	
B		anti-A
AB	A and B	
O		anti-A and anti-B

   d) Complete the gaps in the following table to show which blood transfusions can be carried out safely. A ✓ indicates a safe transfusion.

Patient's blood group	Donor's blood group			
	A	B	AB	O
A	✓			
B		✓	✗	✓
AB		✓	✓	
O	✗			

✓ = no clumping   ✗ = clumping

Examination questions

1 The diagram shows two methods which are used to give humans protection against disease. **Method A** shows active immunity and **Method B** shows passive immunity. **Method A** can be used against polio. **Method B** is often used against tetanus.

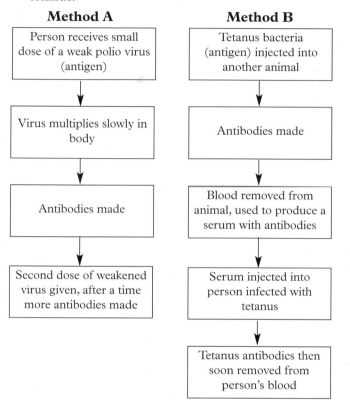

Method A

Person receives small dose of a weak polio virus (antigen)

↓

Virus multiplies slowly in body

↓

Antibodies made

↓

Second dose of weakened virus given, after a time more antibodies made

Method B

Tetanus bacteria (antigen) injected into another animal

↓

Antibodies made

↓

Blood removed from animal, used to produce a serum with antibodies

↓

Serum injected into person infected with tetanus

↓

Tetanus antibodies then soon removed from person's blood

a) What is the name of the substances produced by the body which destroy harmful viruses and bacteria? *(1 mark)*

b) Why does **Method A** give long lasting protection against polio? *(1 mark)*

c) Why does **Method B** not give long lasting protection against tetanus? *(1 mark)*

d) In immunisation against polio a second dose of the weakened virus is given (this is known as a booster). Suggest why this booster is necessary. *(1 mark)*

e) **Method A** would **not** be helpful for a person who had just been infected with tetanus bacteria. Explain the reason for this. *(2 marks)*

f) Why is **Method B** very good for dealing quickly with an infection of tetanus? *(1 mark)*

2 Read the passage.

> ## Could we be on the verge of a modern TB plague?
>
> A dangerous new strain of the bacteria which cause the lung disease, tuberculosis (TB), has recently been found in Britain. Although the victims were treated successfully, doctors fear that the new strain could soon spread. "All it would take would be for someone infected with the new strain to cough in a place like a cinema and the infection could spread like wildfire," said a TB specialist. "No-one is immune to the new strain, and many of the people breathing in the bacteria would develop the disease."

a) Tuberculosis is an infection of the lungs. Most bacteria that we breathe in do not reach the alveoli in the lungs.
Describe how the body prevents bacteria reaching the alveoli. *(2 marks)*

b) The TB specialist says that "No-one is immune to the new strain".
Explain how we can naturally become immune to a disease. *(5 marks)*

c) Many people are immune to the old strain of TB because they have had a vaccine.
　i)　What does a TB vaccine contain that makes a person immune to the disease? *(2 marks)*
　ii)　It is too late to give a vaccine to a person who is already infected with TB.
　　What can be injected to stop the disease developing? *(1 mark)*

3 Injuries, caused by getting a garden fork or rusty nail into the foot, are quite common while gardening.
a) A patient with a deep injury contaminated with soil may be given an injection which contains anti-tetanus antibodies (or anti-serum).
　i)　How will this injection help the patient? *(2 marks)*
　ii)　What is the name given to this type of immunity? *(1 mark)*
b) Vaccination has been very effective in preventing polio. The vaccine currently used in the U.K. contains the polio virus and is given as three drops of liquid in the mouth.
　i)　How is the polio virus treated to make the vaccine safe? *(1 mark)*
　ii)　How can the use of a vaccine containing a virus result in long-lasting immunity? *(1 mark)*

Controlling disease

c) The polio virus enters the body through the mouth and then multiples in the intestine. People travelling to countries where polio is common are given a booster vaccination to prevent them catching the disease.
Suggest the most common way by which they might catch polio if they were not given a booster. *(1 mark)*

4 Colds and influenza ('flu') are infectious diseases caused by viruses. The drawing shows the percentage of people who were affected by colds and flu in different parts of the UK in November 1997.

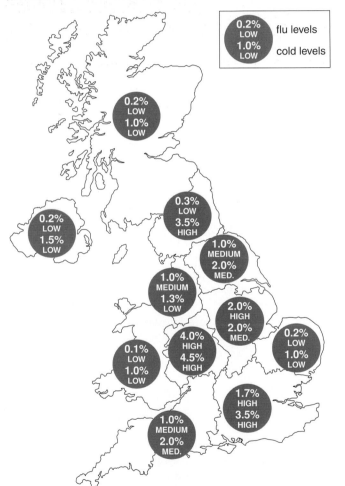

a) What was the highest percentage of people suffering from colds? *(1 mark)*
b) In Scotland only 0.2% of people had flu, compared with 4.0% of people in the Midlands.
   i) The population of Scotland is 5 million. Calculate the number of people in Scotland who suffered from flu in November 1997. Show your working. *(2 marks)*

ii) Suggest an explanation for the much higher percentage of people with flu in the Midlands. *(1 mark)*
c) The symptoms of flu include headache, aching muscles and fever. The treatment is to keep warm and to drink plenty of liquid. Antibiotics are of no use in treating flu.
   i) Suggest **one** medicine that would be useful in treating the symptoms of flu. *(1 mark)*
   ii) Explain how this medicine would be useful. *(1 mark)*
   iii) Explain why antibiotics are of no use in treating flu. *(2 marks)*

5 Tuberculosis (TB) is a disease which usually affects the respiratory system. It has caused more deaths per year world-wide than any other infectious disease. It is caused by a bacterium which can be spread from person to person by coughs and sneezes. The graph shows the changes in earnings and deaths from TB during the period 1850–1940.

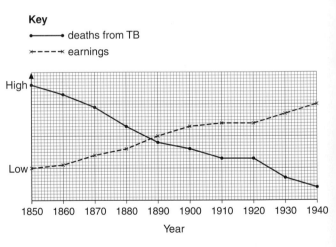

a) i) Why can TB be spread by coughs and sneezes? *(1 mark)*
   ii) What do we call this method of spreading diseases? *(1 mark)*
b) i) Describe carefully the changes in earnings and deaths from TB over the period shown in the graph. *(3 marks)*
   ii) The changes in earnings between 1850 and 1940 changed people's lives in a number of ways. Suggest **two** changes in lifestyle which could have helped to bring about the changes in the number of deaths from TB during this period. *(2 marks)*

6 A patient requires a kidney transplant.
First the transplant doctors must select a kidney from a suitable donor.

a) If the 'tissue-type' of the donor kidney is not matched, the patient's T-lymphocytes may destroy it. Explain how T-lymphocytes may destroy the donor kidney. *(2 marks)*

b) There is a danger that patients receiving transplanted organs may still reject them, even when the donor organ has a 'tissue-type' similar to that of the recipient.
There are three main ways in which the recipient is treated after a transplant:

- irradiating bone marrow,
- drug treatment,
- isolation.

Explain the reason for each one of these treatments. *(4 marks)*

c) During the operation the patient will require a blood transfusion. The patient receiving the transplant has been found to have blood group B.
Complete the table to show the blood groups from a blood donor which can be used for this patient. The table has been completed for a patient with blood group O.

Key: ✓ shows that the blood is compatible
✗ shows that the blood is not compatible

		Donor's Blood Goups			
		A	B	AB	O
Patient's Blood groups	O	✗	✗	✗	✓
	B				

Chapter 10
Using microorganisms

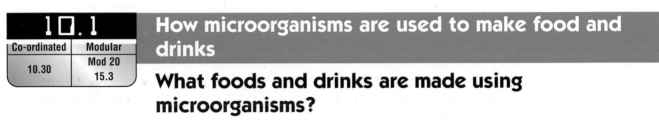

10.1

Co-ordinated	Modular
10.30	Mod 20 15.3

How microorganisms are used to make food and drinks

What foods and drinks are made using microorganisms?

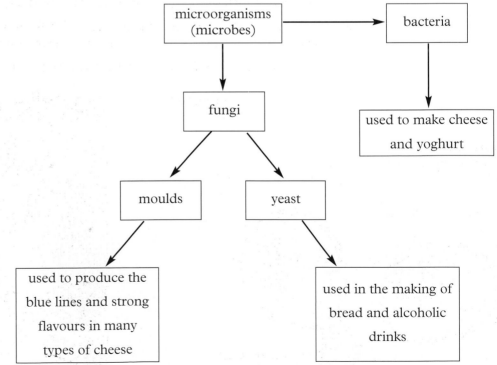

microorganisms (microbes) → bacteria

fungi

bacteria → used to make cheese and yoghurt

fungi → moulds, yeast

moulds → used to produce the blue lines and strong flavours in many types of cheese

yeast → used in the making of bread and alcoholic drinks

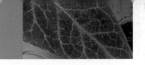

Figure 10.1
The blue veins in these cheeses are caused by mould

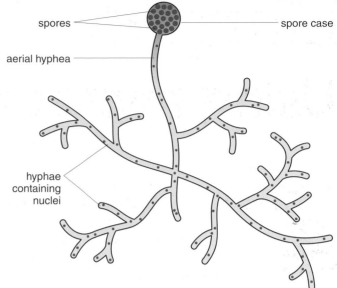

Figure 10.2
The structure of a fungus

Comparing moulds, yeast, bacteria and viruses

- **Moulds** and yeast are fungi. All fungi possess a cell wall, a cell membrane, cytoplasm and distinct organelles, such as a nucleus.

- Moulds are made up of thread-like strands called **hyphae** (singular: hypha). Because the hyphae do not contain separate cells, the cytoplasm contains large numbers of nuclei. Moulds reproduce asexually by producing **spores**. These spores are produced in large numbers. They are very light and can be carried long distances by the wind.

- Yeast is a single-celled fungus that reproduces asexually by budding.

- Bacteria are single-celled organisms. They differ from fungi in that they do not have a nucleus or any other organelles within their cytoplasm.

- Viruses (see section 3.5) consist of a protein coat, surrounding a few genes.

Figure 10.4
Yeast cells budding

Figure 10.3
The structure of a yeast cell

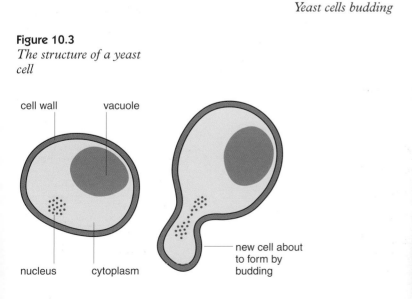

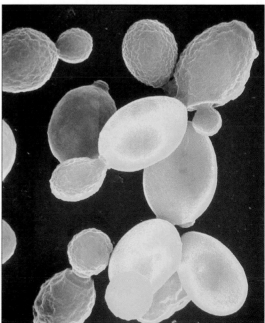

Making use of yeast

Respiration in yeast cells

Yeast can respire without oxygen (**anaerobic respiration**) (see section 2.5). As it respires anaerobically yeast produces carbon dioxide and **ethanol** (alcohol). This process is called **fermentation**.

Yeast can also respire aerobically in the presence of oxygen. This **aerobic respiration** produces carbon dioxide and water. Aerobic respiration provides the yeast cell with more energy than anaerobic respiration does. This is because there is still energy that could be retrieved from the ethanol. The yeast cells use the extra energy they get from aerobic respiration for growth and reproduction.

Yeast and bread-making

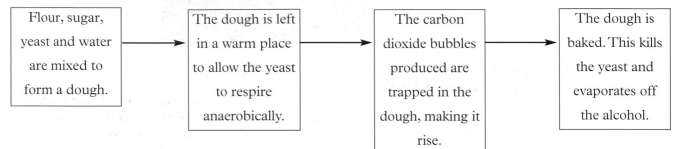

Flour, sugar, yeast and water are mixed to form a dough.	→ The dough is left in a warm place to allow the yeast to respire anaerobically.	→ The carbon dioxide bubbles produced are trapped in the dough, making it rise.	→ The dough is baked. This kills the yeast and evaporates off the alcohol.

Yeast and beer-making

In beer-making the source of energy for the yeast are carbohydrates found in barley seeds (grains).

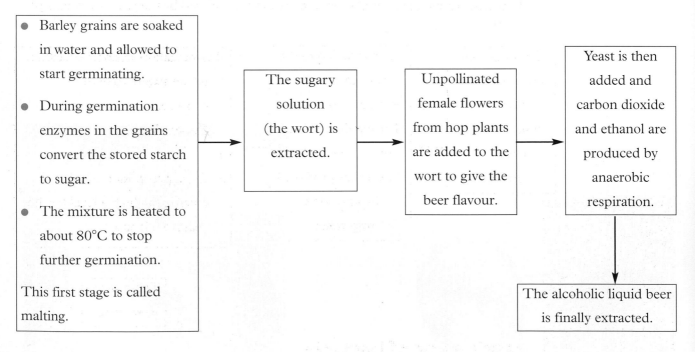

- Barley grains are soaked in water and allowed to start germinating.
- During germination enzymes in the grains convert the stored starch to sugar.
- The mixture is heated to about 80°C to stop further germination.

This first stage is called malting.

→ The sugary solution (the wort) is extracted.

→ Unpollinated female flowers from hop plants are added to the wort to give the beer flavour.

→ Yeast is then added and carbon dioxide and ethanol are produced by anaerobic respiration.

The alcoholic liquid beer is finally extracted.

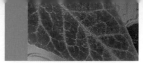

Figure 10.5
A female hop flower

Did you know?

- Today genetically engineered yeasts are often used in beer-making. Some of these produce 'lite' beers, with a low carbohydrate content. Other genetically-engineered yeasts can survive in a high concentration of ethanol, so are used to make stronger beers.

- If female hop flowers get pollinated they acquire a nasty taste. If these flowers are used then the beer could be spoilt.

Yeast and wine-making

In wine-making, the natural sugars in the grapes are used by the yeast as its energy source.

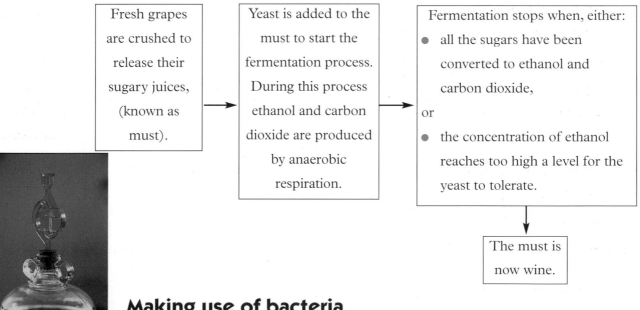

| Fresh grapes are crushed to release their sugary juices, (known as must). | → | Yeast is added to the must to start the fermentation process. During this process ethanol and carbon dioxide are produced by anaerobic respiration. | → | Fermentation stops when, either:
• all the sugars have been converted to ethanol and carbon dioxide,
or
• the concentration of ethanol reaches too high a level for the yeast to tolerate. |

The must is now wine.

Making use of bacteria

Bacteria and yoghurt-making

Yoghurt is made from milk. The advantage of yoghurt over milk is that it lasts longer before going off. The process of making yoghurt depends on providing the correct conditions for the growth of special bacteria (called lactate bacteria) that produce lactic acid as they respire anaerobically.

Figure 10.6
A wine making jar fitted with an air lock

A starter culture of live lactate bacteria are added to warm milk.	→	The bacteria respire anaerobically converting the sugar (lactose) in the milk to lactic acid.	→	The lactic acid causes the milk to clot and solidify into yoghurt.

Bacteria and cheese-making

A starter culture of bacteria is added to warm milk. An enzyme, called rennin, is also added to make the milk curdle faster.	→	Curds are produced that are more solid than those of yoghurt.	→	The curds are separated from the remaining liquid part of the milk (known as whey).	→	Bacteria and moulds are added to the curds to slowly ripen (mature) the cheese.

Did you know?

- Blue cheeses, such as Stilton, are ripened by a mould that can be seen as a network of blue veins that spread throughout the cheese. Before the curd is cut and shaped into a cheese, spores from the mould are added. These germinate and form a crust on the outside. The cheese is then pierced with needles causing the mould to grow inside the cheese. The mould grows inside the cheese along the needle holes forming the characteristic blue veins.

- Some cheeses, such as Gruyere, have large holes in them. These are made during the ripening stage by microorganisms that produce carbon dioxide. This gets trapped in the cheese and makes the holes.

Figure 10.7
Stilton and Gruyere cheeses

Summary

◆ Microorganisms are used to make food and drink.

◆ **Moulds** and yeast are fungi.

◆ Yeast is a single-celled organism.

◆ Moulds have thread-like structures called **hyphae**.

◆ Hyphae are not divided into separate cells.

◆ Moulds reproduce asexually by producing spores.

◆ Yeast can respire anaerobically producing ethanol and carbon dioxide. This is called **fermentation**.

◆ Yeast is used in bread-making, in the brewing of beer and in wine-making. In each fermentation process different carbohydrates are used as the energy source.

◆ Bacteria are used in the production of yoghurt.

◆ Bacteria and moulds are used in the production of various cheeses.

Topic questions

1 Which microorganisms are used in the making of:
   a) cheese?
   b) yoghurt?
   c) bread?
   d) beer?
   e) wine?

2 a) In what ways are bacteria and fungi similar?
   b) In what way are they different?

3 a) What is yeast?
   b) How does it reproduce?

4 Moulds are made up of hyphae. What are hyphae?

5 When yeast respires anaerobically what two substances are produced?

6 What causes bread to rise?

7 What provides the energy for the yeast in:
   a) beer-making?
   b) wine-making?

8 What acid is made by the bacteria used in the production of yoghurt?

10.2   Other useful substances made using microorganisms

Co-ordinated	Modular
10.31	Mod 20 15.4

Many useful substances, such as penicillin and insulin, can be extracted from microorganisms. To obtain a sufficient quantity of these substances, very large numbers of microorganisms need to be grown. This is carried out in an **industrial fermenter**. These are units that are specially designed to provide an environment in which the temperature, pH and nutrients required by the particular microorganism can be monitored and carefully controlled.

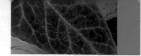

Using microorganisms

A typical industrial fermenter consists of the following parts:

- a *vessel*, usually made of stainless steel, filled with the microorganism and a medium containing the required nutrients;

- a *stirrer*, to keep the microorganisms in suspension and to maintain an even temperature throughout the medium;

- an *air supply* (aerator) to provide oxygen for the microorganisms to respire aerobically;

- a *water-cooled jacket* to remove the heat produced by the fermenting microorganisms;

- *probes* to monitor pH and temperature.

Figure 10.8
An industrial fermenter used to produce antibodies for vaccines

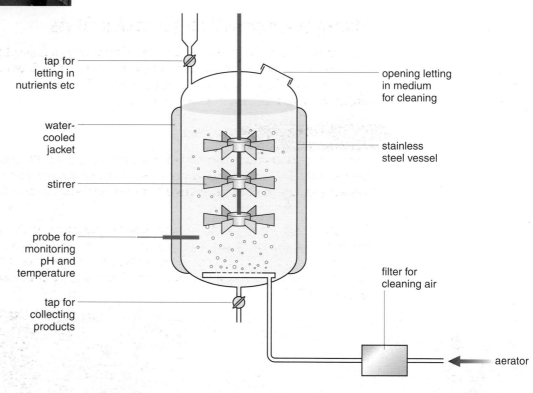

Figure 10.9
A model of an industrial fermenter

It is essential that the container, the nutrients and any air supply are sterilised before use. This is to prevent contamination by unwanted microorganisms.

When fermentation is finished, the contents of the vessel are drained off and the useful products are separated out and purified.

Making penicillin

The antibiotic penicillin is grown from the mould *Penicillium*, see section 9.1. If this is provided with large amounts of all the nutrients it needs it will grow rapidly, but will not produce large amounts of penicillin. Penicillin is only produced in large amounts if the mould is competing for the food available with another organism, such as a colony of bacteria, or if the food supply is limited. In the industrial fermenter the amount of food is deliberately limited, so the *Penicillium* reacts by producing large amounts of penicillin – most penicillin being produced when the nutrients have been used up.

Figure 10.10
Graph to show the yield of penicillin over time

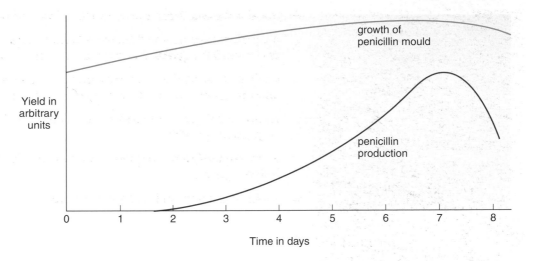

Using fermentation to make fuels

Methane, (CH$_4$), the main constituent of natural gas, is produced when organic material, such as sewage, manure, or dead plant material, decompose anaerobically by the action of a variety of microorganisms. Methane also forms inside the digestive system of cows (see section 6.2).

> **Did you know?**
>
> The action of the digestive bacteria in a cow can release up to 500 litres of methane in a 24-hour period. It is estimated that if this gas could be collected from just three cows, there would be enough to supply an average home with all the gas it needs for 24 hours.

The gas produced by decomposing organic material is called biogas. It consists mainly of methane. To make biogas, the waste organic material is put into a biogas generator (also called a biogas digester) where fermentation of the carbohydrates present in the material takes place. The biogas can be collected and used as an energy source.

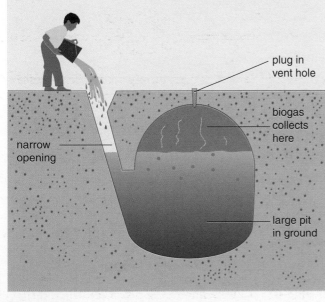

Figure 10.11
A biogas generator

Figure 10.12
The waste from this toilet flows into the biogas generator in the gound next to it

193

It is important that the biogas generator is built in such a way as to restrict the amount of air that can enter. This will enable the anaerobic conditions needed for the fermentation to be maintained. If too much oxygen becomes available then complete oxidation will occur producing only carbon dioxide and water.

Making use of the biogas

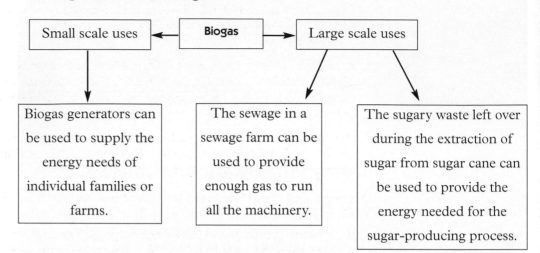

| Small scale uses | ← Biogas → | Large scale uses |

Biogas generators can be used to supply the energy needs of individual families or farms.

The sewage in a sewage farm can be used to provide enough gas to run all the machinery.

The sugary waste left over during the extraction of sugar from sugar cane can be used to provide the energy needed for the sugar-producing process.

Using anaerobic fermentation to produce fuels for motor vehicles

Ethanol-based fuels can be made by the anaerobic fermentation of:

- sugar cane juices, which are produced during the sugar-making process

- glucose, which is obtained by the action of the enzyme carbohydrase on maize starch.

The products of fermentation are distilled to produce the ethanol that, when mixed with petrol, can be used as a fuel for motor vehicles.

Did you know?

- In China, the fermentation of animal waste has been used to produce biogas for many years. Today, there are 10 million biogas generators in use.

- In Brazil, over 6 million vehicles are run on an ethanol-based fuel.

The economic and environmental advantages and disadvantages of the production of fuels by fermentation and their use

There are two reasons why alternatives to fossil fuels as energy sources need to be developed. These are:

- the burning of fossil fuels is the main contributor to global warming;

- it is predicted that the reserves of fossil fuels will only last another 40 to 50 years.

One alternative is the development of a fuel that is obtained from organic matter by the anaerobic fermentation of such organic materials as sugar cane, animal waste, household organic waste and sewage.

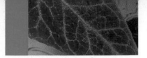

The advantages of developing and using such fuels are that they:

- are CO_2 neutral. This means that the carbon dioxide given out when the fuel is burned is balanced by the carbon dioxide taken in whilst the crops are growing.

- make use of waste organic material, the disposal of which often creates environmental problems. Such waste can include animal waste, especially from those animals reared in confined spaces. These generate large amounts of waste in a small area. Manure from pigs, cattle and chickens is converted by anaerobic fermentation into biogas. The biogas can be used as a fuel for generators used to produce electricity.

- are a renewable energy source that makes use of unsophisticated technology. In countries such as India, China and other Far East countries the development of such fuels can satisfy the energy requirements of a large proportion of the population.

- provide a variety of fuels. The fuel could be a gas that is burned or a liquid that can be used as a fuel for vehicles. Other forms of renewable energy, such as tides, wind or solar power are only used to generate electricity.

The main disadvantages are concerned with the cost and the availability of land on which crops can be grown.

- If crops are grown just to provide the organic material for the fermentation process, then there could be less land available for the growing of crops for food production.

- The cost of fuels produced by the fermentation process is, at the moment, more expensive than the cost of fossil fuels. This is because the present low price of fossil fuels does not take into account the exploring, extracting and refining costs.

Summary

- Large numbers of microorganisms can be grown in **industrial fermenters**.

- These consist of an air supply, stirrer, water-cooled jacket and various monitoring probes.

- Penicillin is made from the mould *Penicillium*.

- The mould produces most penicillin when the food supply is low.

- Fuels can be made from organic materials by fermentation.

- **Biogas**, which is mainly methane, is made by the anaerobic fermentation of plant products or waste materials containing carbohydrates.

- **Ethanol**-based fuels can be produced by the anaerobic fermentation of sugar cane juices, and from glucose produced by the action of carbohydrase on maize starch.

- The ethanol-based fuels can be used in motor vehicles as a fuel.

Topic questions

1 a) What is the function of an industrial fermenter?
   b) What are the five main parts of an industrial fermenter? What is the purpose of each part?
   c) Penicillin can be made in an industrial fermenter.
       i)   From what mould is penicillin extracted?
       ii) Under what conditions does this mould produce most penicillin?

2 a) What is biogas?
   b) How is biogas produced?
   c) Why should a biogas generator be designed to limit the amount of oxygen that is available?

3 Give two sources of carbohydrates that are fermented industrially to produce ethanol.

4 a) Explain three advantages that fuels fermented from carbohydrates have over fossil fuels.
   b) Give one disadvantage that fuels fermented from carbohydrates have over fossil fuels.

10.3 · Using microorganisms safely

Co-ordinated	Modular
10.30	Mod 20 15.5

Providing the food supply

Bacteria and other microorganisms cannot make their own food. If they are to be grown they therefore need a supply of food. The food should contain an energy source, which will usually be carbohydrates, together with proteins, mineral ions and vitamins. In the laboratory the colonies (cultures) of the microorganisms are normally grown on **agar** jelly which contains all the nutrients required for growth.

Did you know?

- Agar is an extract from seaweed, to which the required nutrients for the growth and reproduction of the microorganisms are added.

- At high temperatures agar is a liquid, but at room temperature it sets into a jelly. It is poured into a petri dish to form an agar plate and allowed to cool.

- When bacteria, or other microorganisms, are introduced onto the surface of the solidified agar they grow and reproduce to produce visible colonies.

Figure 10.13
Colonies of bacteria growing on agar

Preparing uncontaminated cultures

Many of the cultures of microorganisms grown on agar are used to provide useful products, such as penicillin. It is therefore essential that only the required microorganism grows, and that the culture is uncontaminated with microorganisms from another source.

In order to ensure these conditions:

- the petri dishes and culture medium (the nutrient agar), must be sterilised before use to kill any unwanted microorganisms;

- the nutrient agar, once sterilised, must be covered until required;

- the wire inoculating loops used to transfer the required microorganisms to the nutrient agar must be sterilised by passing them through a Bunsen flame until they are red hot;

- the lid of the petri dish must be sealed with adhesive tape to prevent any microorganisms from the air contaminating the culture.

In school and college laboratories cultures should be incubated at temperatures no greater than 25°C. This is to prevent the growth of pathogens that might be harmful to humans. Higher temperatures than this are permitted in industrial processes because the higher temperatures result in a more rapid growth rate.

Summary

- Large numbers of microorganisms can be grown in a culture medium.

- The culture medium needs to supply the microorganisms with all the nutrients required for growth.

- These nutrients are often contained in **agar** jelly.

- The production of uncontaminated cultures requires that all equipment involved should be sterilised.

- The containers must be sealed.

- Incubation temperatures must not exceed 25°C in schools and colleges.

- Higher temperatures can be used in industry to produce more rapid growth rates.

Topic questions

1   a)   What nutrients should be provided for a growing colony of bacteria?
    b)   What is the name of the jelly on which microorganisms are grown?

2   a)   Describe the stages involved in the making of an uncontaminated culture of a certain bacterium.
    b)   Why is it important to make sure that a colony of bacteria is not contaminated?

3   In a school laboratory microorganisms must not be cultured at temperatures higher than 25°C. Why?

Examination questions

1 Yeast is a fungus that will reproduce in sugar solution.
   a) A student stained some yeast cells and looked at them using a microscope. The majority appeared as shown in **diagram A**.

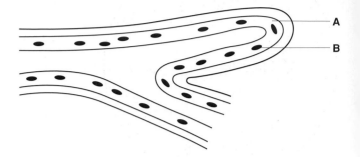

Diagram A

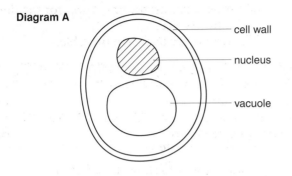

cell wall

nucleus

vacuole

Some cells appeared as shown in the following **diagram B**.

Diagram B

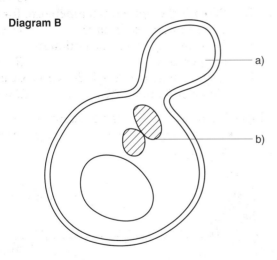

a)

b)

Copy **diagram B** and label A and B to explain what process is taking place in this cell.
   (2 marks)

2 Microbes can be used to make some medicines.
   a) Name **three** other useful products which microbes can make for us. *(3 marks)*
   b) The drawing shows part of a hypha from a type of microbe.

   i) On the drawing, use words from the list to name the parts labelled **A** and **B**.
   (2 marks)

   **chloroplast   cytoplasm   membrane
   nucleus   wall**

   ii) What type of microbe does the drawing show? *(1 mark)*
   When using microbes to make medicines, the microbes are grown in a culture medium which must contain certain substances. One of these substances is carbohydrate.
   c) i) Why must the culture medium contain carbohydrate? *(1 mark)*
      ii) As well as carbohydrate, the culture medium also contains water. Name **two** other substances which could be needed in the culture medium. *(2 marks)*
   Sometimes the microbes are grown on a jelly-like medium in a petri dish.
   d) Why must the petri dish and the medium be sterilised before being used? *(1 mark)*
   e) A wire inoculating loop is used to transfer the microbes to the petri dish. How could the inoculating loop be sterilised? *(1 mark)*
   f) When making medicines, how could the microbes be grown as quickly as possible in the petri dish? *(1 mark)*

3 A new disease was found to be caused by a type of bacterium. An investigation was carried out to find out which antibiotics were effective against this type of bacterium. Each of four paper discs, **A** to **D**, were soaked in a different antibiotic. A fifth disc, **E**, was a control.
   A culture containing the bacteria was used to inoculate the surface of the growth medium in a petri dish. The five paper discs were put on the surface, and the dish was incubated for 24 hours. The drawing shows the result of this investigation.

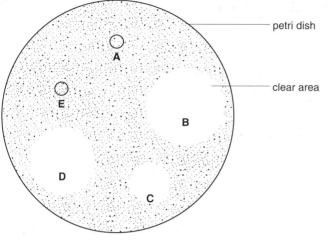

a) How should the control disc, **E**, be treated before putting it into the petri dish? *(1 mark)*

b) Explain, as fully as you can, why there were clear areas around some discs after 24 hours. *(2 marks)*

c) From the results shown in the diagram, what is the order of effectiveness of antibiotics **A** to **D** against this type of bacterium? *(1 mark)*

The antibiotic, streptomycin, is produced by a mould which is grown in a fermenter. The graph shows the concentrations of mould, glucose and streptomycin in a fermenter over a period of ten days.

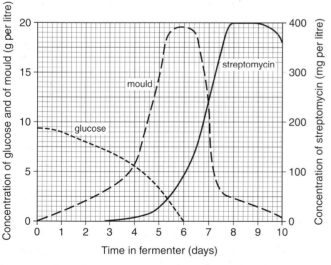

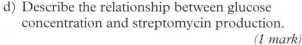

d) Describe the relationship between glucose concentration and streptomycin production. *(1 mark)*

e) Explain
  i)   the change in glucose concentration; *(2 marks)*
  ii) the decrease in concentration of the mould. *(2 marks)*

4   A mixture of yeast, sugar, flour and water is used to make bread dough. After mixing, the dough begins to rise.
   a) Explain why the dough rises. *(3 marks)*
   b) A baker wants the dough to double in size before baking. The baker investigated the effect of temperature on the time taken for the dough to double its size.
   The results are shown in the table.

Temperature (°C)	0	10	20	30	40	50	60
time taken for dough to double in size (minutes)	did not rise	72	48	23	16	37	did not rise

   Draw a graph to show the effect of temperature between 10°C and 50°C on the time taken for the dough to double in size. Plot temperature (°C) on the x axis and time taken for dough to double in size (minutes) on the y axis. *(3 marks)*
   c) i)   What is the optimum temperature for the dough to double in size? *(1 mark)*
      ii) Explain why this is the optimum temperature. *(2 marks)*
      iii) Explain why the dough does not rise at either 60°C or at 0°C. *(3 marks)*

5   Biogas is a useful fuel. It can be made by microbes. The diagram shows one design for a biogas generator.

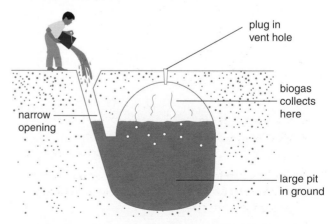

a) Suggest **two** sources of waste material which could be put into the biogas generator to produce biogas. *(2 marks)*

b) What is the main gas in biogas? *(1 mark)*

c) Suggest advantages of having a narrow opening to the generator. *(3 marks)*

6 The diagram shows some stages in making cheese.

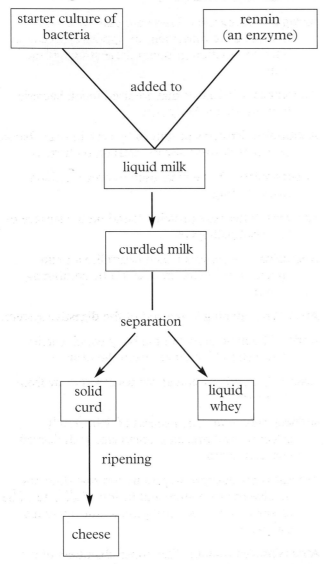

a) The bacteria in the starter culture ferment the sugar in the milk.

i) Name the acid produced. *(1 mark)*

ii) What is the advantage to the bacteria of fermenting the sugar? *(1 mark)*

The enzyme, rennin, was added to make the milk curdle more quickly. An investigation was carried out to find the effect of pH on the activity of rennin. The results are shown in the table.

pH	2	3	4	5	6
time taken to curdle milk (minutes)	4	2	3	6	13

b) Draw a graph of these results with pH on the x axis and time taken to curdle milk (minutes) on the y axix. *(2 marks)*

c) What was the best pH for the rennin to curdle the milk? *(1 mark)*

d) At pH 8 rennin does not make milk curdle. Suggest why rennin does not curdle milk at pH 8. *(1 mark)*

Glossary

Absorption (digestion) The process in digestion whereby water or small soluble particles pass through the lining of the intestine into the bloodstream.

Absorption (plants) The process whereby plants take in water or dissolved mineral ions through their root hairs.

Active immunity The acquiring of immunity against a disease through being given small quantities of dead or inactive forms of the pathogen.

Active transport The process by which cells take up substances against a concentration gradient. The process requires energy from respiration.

Adaptation A feature of an organism that helps it survive in a particular environment.

Addiction Being unable to do without a drug such as nicotine, alcohol or heroin.

ADH (anti-diuretic hormone) A hormone released from the pituitary gland which targets the kidney tubules to ensure water is reabsorbed into the blood so reducing the amount of water in urine and controlling water balance.

Aerobic respiration Respiration that takes place in the presence of oxygen.

Aerofoil When air flows around this shape it produces lift.

Agar The jelly to which nutrients are added, and on which colonies of micro-organisms are grown.

Agglutination The clumping of the blood that can occur during a blood transfusion if the same type of antigen and antibody are present in the donor's and recipient's blood.

Air A mixture of gases made up of approximately 4/5 nitrogen and 1/5 oxygen.

Alleles One of the different forms of a particular gene. For example the allele for eye colour can be for blue or brown eyes.

Alveoli (singular: alveolus) A microscopic air sac in the lungs which acts as the surface for gaseous exchange.

Amino acids The breakdown products of the digestion of proteins, and the building blocks for making new proteins.

Amylase An enzyme that digests starch into sugar.

Anaerobic respiration Respiration that takes place in the absence of oxygen.

Antagonistic muscles These are pairs of muscles that produce movement in opposite directions. The contraction of one muscle stretches the other.

Antibiotics Chemicals that are used to kill bacteria that invade the body.

Antibodies Proteins produced by certain white blood cells that destroy disease-causing pathogens.

Anticoagulant A chemical used to prevent blood from clotting.

Antigens These are proteins found on the surface of bacterial pathogens.

Antitoxins Substances which neutralise a toxin (poison) produced by bacteria by combining with it.

Anus The opening at the end of the digestive system.

Aorta The main artery of the body which carries oxygenated blood away from the heart.

Artery A blood vessel that carries blood away from the heart.

Artificial selection The process of deliberately selecting and breeding organisms with desired characteristics.

Asexual reproduction Reproduction that does not involve the formation and fusion of gametes. The offspring have identical genetic information to the parent.

Atria (singular: atrium) The upper chambers of the heart which receive blood from the veins.

B cells These are lymphocytes that produce antibodies.

Bacteria (singular, bacterium) A single celled organism consisting of cytoplasm and a membrane surrounded by a cell wall, with genes not organised to form a distinct nucleus.

Baleen This is the material used as a sieve in the jaws of filter feeding whales. It is a strong flexible material made of the same protein as is present in human fingernails.

Glossary

Bile A liquid produced by the gall bladder that breaks up fats into droplets.

Biodegradable Materials which can be broken down (decomposed) by bacteria.

Biogas The gas, mainly methane, produced when organic material decomposes anaerobically.

Biogas generator A device in which organic material is allowed to decompose anaerobically.

Biomass The dry mass of living material in an ecosystem.

Bladder The sac which fills with urine from the kidney.

Brain The part of the central nervous system which controls and co-ordinates most of the body's activities.

Breathing The action of passing air into and out of the lungs.

Caecum This is a bag which opens into the digestive system at the junction of the small and large intestines.

Calcium phosphate This is the main calcium salt in bones that gives a bone its hardness.

Cancer A group of cells that are dividing very much more rapidly than is normal.

Capillaries Fine, thin-walled blood vessels which form a network for the exchange of substances with the tissue cells.

Carbohydrase An enzyme that can be used to convert starch to glucose.

Carbon dioxide A gas formed during respiration and in the combustion of hydrocarbons. It turns clear limewater milky.

Carnassial teeth These are the enlarged premolar teeth found in the jaws of a carnivore, such as a dog, and which are used in a shearing action to cut flesh from their prey.

Carnivore An animal obtaining its food as meat from other animals.

Cartilage This is the smooth layer which covers the ends of the bones and which stops the bones rubbing together.

Catalyst A substance that changes the speed of a reaction but remains unchanged after the reaction.

Cell The smallest part of an animal or plant.

Cellulose The material that makes up the cell wall of plant cells.

Cell division The processes of mitosis and meiosis.

Cell membrane The very thin membrane on the outside of a cell that controls the movement of substances in and out of the cell.

Cell wall The outer part of a plant cell that gives strength and shape to the plant cell.

Characteristic A feature of an organism that can be observed.

Chlorophyll A green pigment found in the leaves and stems of plants. It traps light energy for use in photosynthesis.

Chloroplast That part of the plant cell containing chlorophyll.

Chromosome One of the thread-like structures found in the nucleus which contains genetic material. They are made of a single very long molecule of DNA. In humans there are 23 pairs of chromosomes in each body cell, but 23 single chromosomes in each gamete.

Ciliary muscles Muscles in the eye which control the thickness of the lens when focussing on near and distant objects.

Clone An organism produced asexually from one parent. The clone will be genetically identical to the parent.

Competition The interaction of organisms which are trying to obtain the same food or occupy the same space.

Compression The process of being squashed.

Concentration gradient This exists wherever there is a difference in the concentration of a substance in two areas.

Connector neurone (relay neurone) Neurones found in the spinal cord which link sensory and motor neurones.

Consumer An organism which has to rely on food made by green plants (producers) or other animals.

Cornea The transparent front part of the eye. Plays an important part in focussing light onto the retina.

Cytoplasm All the material in a cell inside the membrane (apart from the nucleus), where chemical reactions take place under the control of enzymes.

Glossary

Decay The breaking down by decomposers of complex organic materials into simple ones.

Decomposers Organisms which break down complex organic materials into simple ones during decay. Most are bacteria and fungi.

Deforestation The removal of trees from woodland and mountain sides. It often leads to flooding of rivers as the trees can no longer take up the rain that falls on them.

Denaturing The process by which enzymes are destroyed when heated above a temperature of about 40 °C

Denitrifying bacteria Bacteria which convert nitrates in the soil into nitrogen gas.

Dental formula The numbers of each type of tooth in a jaw.

Dentition The types of teeth and their arrangement in the mouth.

Deoxygenated blood Blood that is not rich in oxygen.

Detritivores Organisms, such as termites, earthworms, fungi and bacteria that obtain energy and nutrients by feeding on dead organic matter. The decomposers are a class of detritivores.

Diabetes A disease caused by lack of insulin production from the pancreas. Blood sugar levels cannot be controlled properly.

Dialysis The treatment of kidney failure by taking blood from the patient and removing the urea and other waste products by diffusion.

Dialysis machine An artificial kidney machine.

Diaphragm A sheet of muscle which separates the thorax from the abdomen. Flattening of the diaphragm results in air entering the lungs.

Diastema The space between the premolars and the incisors in the jaw of a herbivore such as a sheep.

Diffusion The movement of particles resulting in a net movement from a region where they are at a high concentration to a region where they are at a lower concentration.

Digestion The process by which food is broken down into particles small enough to be absorbed into the blood.

Dilate To get wider.

DNA The chemical that carries the genetic information on the genes.

Dominant The allele that controls an observable characteristic (phenotype) in the offspring even when it is present on only one chromosome (heterozygous).

Donor The person who, in a kidney transplant, provides the replacement kidney.

Ecosystem The organisms living and surviving in a particular place.

Effector neurone Nerve cell which carries impulses away from the spinal cord towards effectors.

Effectors Structures such as muscles or glands which carry out responses to stimuli.

Emulsifying Breaking down of a liquid into very fine droplets.

Endocrine gland A gland which discharges its products, called hormones, straight into the blood.

Endoskeleton The hard bony skeleton of vertebrates, that forms and grows inside the body.

Environment The surroundings and conditions that affect the growth and behaviour of plants and animals.

Enzyme A protein that can act as a catalyst for a reaction. It can be easily destroyed (denatured) by heating.

Ethanol The alcohol produced by the anaerobic respiration of yeast.

Eutrophication A process caused when large amounts of nitrates and phosphates are discharged into rivers and streams. The nutrients cause the rapid growth of algae and water plants. The eventual death of the algae and plants soon leads to the rapid growth of aerobic bacteria. These decomposers soon use up all the available oxygen in the water. This in turn causes other animal life in the water to suffocate and die.

Excretion The removal of chemical waste material made in the body or a cell.

Exhale To breathe out.

Exoskeleton The hard, outside covering on the bodies of many crustaceans and insects, which provides protection and support for all the internal organs.

Extinct A description of an organism no longer living today but which according to the fossil record has lived in the past.

Eye A sense organ that contains the receptors sensitive to light.

Faeces The remains of the undigested food that is passed out through the anus.

Fatty acids The breakdown products of fats.

Fermentation The changing of glucose into ethanol (alcohol) and carbon dioxide by the action of enzymes in yeast.

Fertile The ability of a male or female to produce sex cells which are capable of producing viable offspring.

Fertilisation The fusion of an egg and a sperm.

Fertiliser A substance which can be natural or artificial applied to soil to improve the growth of plants.

Fertility drugs Drugs that stimulate the release of eggs from the ovaries.

Filter feeders These are animals that obtain their food by sieving the water and trapping any plankton it contains.

Filtration Filtration helped by the high pressure of the blood in the capillaries in the glomerulus resulting in the first stage of the formation of urine.

Filtration (kidney) Filtration of water and ions from the bloodstream, to form urine.

Fluid feeders These are animals that need liquid food.

Focus The formation of a sharp image of near and distant objects by altering the shape of the lens.

Food chain A diagram which shows feeding relationships of some organisms in an ecosystem. All food chains start with producers which trap light energy.

Fossil The remains or imprints of dead plants or animals trapped in sedimentary rocks when the rocks were formed. The remains or imprints may have been mineralised and turned into stone.

Fossil fuels The non-renewable energy resources: crude oil, natural gas and coal.

Fossilisation The process that produces fossils.

FSH (follicle-stimulating hormone) The hormone secreted by the pituitary gland that causes eggs to mature and stimulates the ovaries to produce oestrogen.

Fusion The process that occurs when the nucleus of a male gamete combines with the nucleus of a female gamete.

Gall bladder A small organ joined to the liver that stores bile.

Gamete A sex cell.

Gaseous exchange Occurs in the alveoli in the lungs when oxygen diffuses across the alveolar membrane from the lungs to the capillaries and carbon dioxide diffuses from the blood capillaries into the alveoli.

Gene A unit of inheritance controlling one particular characteristic and made up of a length of chromosomal DNA.

Genetic Related to inheritance.

Genetic engineering The deliberate changing of the characteristics of an organism by manipulating chromosomal DNA.

Genotype The genetic make-up of an individual. The sum total of all the genes even if they are not shown in the individual.

Gland A structure that releases hormones into the bloodstream.

Global warming An international problem caused partly by the increase in the amounts of carbon dioxide and methane in the atmosphere which results in an increase in the average temperature of the Earth.

Glucagon A hormone released by the pancreas that causes the liver to convert glycogen into glucose.

Glycerol A breakdown product from the digestion of fats and oils.

Glycogen The form in which excess glucose in the blood is stored in the liver and muscles.

Greenhouse effect The effect in the atmosphere of heat energy being absorbed by gases such as carbon dioxide and methane.

Guard cells Pairs of cells which surround the stomata on the surface of leaves which by means of osmosis open and close the stomata thus regulating the flow of gases into and out of the leaf.

Gullet *see oesophagus.*

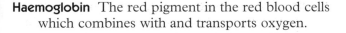

Haemoglobin The red pigment in the red blood cells which combines with and transports oxygen.

Heart A double pump with the right side pumping blood at low pressure to the lungs to release carbon dioxide and collect oxygen and the left side pumping oxygenated blood at higher pressure around the body.

Herbicide A chemical used to destroy unwanted plants.

Herbivore An organism that feeds only on plants.

Heterozygous The inheriting of one dominant allele and one recessive allele for a particular characteristic.

Homeostasis The automatic control system by which the internal conditions of an organism are kept steady.

Homozygous The inheriting of two dominant or two recessive alleles for a particular characteristic.

Hormone A substance secreted by endocrine glands directly into the blood in one part of the body and carried in the blood plasma to a target organ. Plant hormones are called auxins.

Hyphae The thread-like parts of a mould.

Image The picture formed on the retina of the eye.

Immunisation (see vaccination) The introduction of small quantities of dead or inactive forms of the pathogen into the body.

Industrial fermenter The device in which large numbers of particular micro-organisms are grown commercially.

Inhale To breathe in.

Insoluble A substance that will not dissolve in a liquid, usually water.

Insulin A hormone released by the pancreas which helps to control sugar level in the blood.

Insulin controls the conversion of excess glucose into glycogen which is then stored in the liver and muscle cells.

Iris A ring of muscle which controls the amount of light entering the eye.

Joints These are formed wherever two or more bones come together.

Kidney An organ which removes excess water from the blood and excretes urine made from the urea produced in the liver.

Kidney transplant An operation in which a person with failing kidneys receives a healthy kidney from someone else.

Lactic acid A product of anaerobic respiration in very active human muscles which is a mild tissue poison (causes the muscles to hurt).

Large intestine The part of the digestive system where water is removed from indigestible food.

Lens A transparent structure within the eye that is flexible and helps light to form a sharp image on the retina during focussing.

LH (luteinising hormone) The hormone secreted by the pituitary gland that stimulates the release of an egg.

Lift The upward force created by an aerofoil.

Ligaments These are strong fibres that hold bones firmly together.

Limiting factor The factor such as light intensity (brightness), light wavelength, water and carbon dioxide that limits the rate of photosynthesis at a given time.

Lipids Foods made up of fats and oils.

Lipase An enzyme that digests fats to fatty acids.

Liver An organ where excess glucose in the blood is stored as glycogen, where bile is produced and where poisons such as alcohol are removed from the blood.

Locomotion Locomotion is the ability of a living organism to move its body from one place to another.

Lymphocytes Types of white blood cells.

Matrix This is a solid material which cells secrete around them so that the cells are pushed apart leading to cells being scattered in the matrix.

Median fins These are dorsal and ventral fins on a fish. They keep the fish upright.

Meiosis Cell division that leads to the production of gametes in which there has been some reassortment of genetic material so producing variation. It is a reduction division so each gamete has only half the number of chromosomes as the parent.

Memory cells Cells that remain in the body after an infection so that antibodies can be made rapidly if the same infection reoccurs.

Glossary

Migration The mass movement of organisms on a regular basis. Most migrations are connected with seasonal changes and enable organisms to maintain food supplies.

Mitochondria (singular: mitochondrion) The parts of the cell in which aerobic respiration takes place producing cellular energy.

Mitosis Cell division that occurs during growth and asexual reproduction and involves each chromosome making an exact copy of itself, resulting in the formation of two daughter cells each with the same number of chromosomes as the parent.

Motor neurones Neurones that carry electrical impulses from the brain or spinal cord to an effector.

Mould A type of fungus.

Mucus A sticky fluid which traps dust or protects surfaces.

Muscle fibres These are the individual fibres that make up muscle tissue.

Mutation A change suddenly occurring in one or more of the genes or chromosomes or in the number of chromosomes.

Mutualism or symbiosis This is the relationship between two organisms, usually of different species, in which both gain a benefit from living together.

Myotomes These are blocks of muscles.

Natural immunity The acquiring of immunity when the body produces antibodies in response to an infection.

Natural selection The process by which beneficial characteristics with greater survival value are selected and increase in proportion in the population. Natural selection leads to evolution.

Negative feedback An automatic control mechanism in which a change from the normal condition triggers off a response which restores the normal condition.

Nerve impulses Electrical signals which travel along nerve pathways made up of nerve cells (neurones).

Neurone A cell in the nervous system.

Nicotine The addictive substance in tobacco.

Nitrates Chemicals containing NO_3 ions, frequently used in fertilisers to help plants synthesise proteins.

Nitrifying bacteria Bacteria which convert ammonium compounds in the soil into nitrates.

Nucleus (cells) The part of a cell that contains the chromosomes which carry the genes controlling the cell's characteristics.

Nutrition The process by which organisms obtain their raw materials and absorb useful substances from it.

Oesophagus (gullet) The muscular tube which carries food from the mouth to the stomach.

Oestrogen A hormone produced by the ovaries which controls female sexual characteristics.

It inhibits the production of FSH and stimulates the release of LH.

Omnivore An animal obtaining its food from both animal and plant sources.

Optic nerve A bundle of nerve cells which carries impulses from the eye to the brain.

Oral contraceptive Tablets, usually containing oestrogen, that inhibit the production of FSH so that no eggs mature.

Organ A group of tissues working together to carry out a particular function.

Organ system A group of organs working together to carry out a particular function or group of related functions.

Organic Compounds of carbon found in large quantities in living and dead organisms.

Organism An individual plant or animal.

Osmosis The diffusion of water through a partially-permeable membrane – the water flowing from a region of high water concentration to a region of lower water concentration.

Oxygen The chemical element that is vital to life. It will relight a glowing spill.

Oxygenated blood Blood rich in oxygen.

Oxygen debt The oxygen needed to remove the lactic acid from the muscles produced as a result of muscles respiring anaerobically during vigorous exercise.

Oxyhaemoglobin The chemical formed when oxygen combines with haemoglobin.

Paired fins These are the pectoral and pelvic fins on a fish. They are found on both sides of the body and help the fish to move upwards and downwards.

Palisade cells The cells in the upper part of green leaves which contain most chlorophyll and carry out most of the photosynthesis in the leaf.

Pancreas An organ of the digestive system that produces the hormone insulin and the enzyme lipase.

Parasite This is an organism that lives on or in another organism and obtains all its food from it.

Partially permeable membrane Membrane that allows small molecules to pass through quickly, but not large molecules.

Passive immunity The acquiring of immunity against a disease through being given antibodies to counteract the effects of a particular pathogen.

Pathogens These are the micro-organisms that cause disease.

Penicillin The first antibiotic to be used.

Pentadactyl limb This is the basic arrangement of limb bones in vertebrates.

Pesticide A chemical designed to kill unwanted organisms.

pH A scale used to measure acidity and alkalinity.

Phenotype The way an individual appears as a result of the alleles it carries and the environment in which it has grown up.

Phloem A column of cells in a plant responsible for the transport of food made in photosynthesis to wherever it is needed.

Phosphates Chemicals containing PO_4 ions, frequently used as fertilisers to help plants photosynthesise and respire.

Photosynthesis The process in green plants which produces biomass (initially carbohydrates) and oxygen, and requires carbon dioxide and water as raw materials and chlorophyll to enable the plant to absorb light energy.

Pituitary gland A gland, found at the base of the brain, that secretes FSH and LH.

Plankton These are the very small plants and animals that live in water.

Plasma The straw-coloured liquid part of the blood which transports cells and dissolved substances.

Platelets Cell fragments which help in forming blood clots at wounds.

Pollution The introduction of harmful substances into an environment.

Population The numbers of one species of animal living in a particular area.

Potassium An element used by plants to help the action of the enzymes involved in photosynthesis and respiration.

Predation The process by which one animal (**predator**) catches then eats another (**prey**).

Primary feathers The flight feathers fixed to the ulna bone. They form the aerofoil section of the wing.

Producer A green plant which photosynthesises to make its own food.

Proteases Enzymes that digest proteins into amino acids.

Protein fibres These help to make bones hard.

Pulmonary artery The blood vessel that takes deoxygenated blood from the heart to the lungs.

Pulmonary vein The blood vessel that takes oxygenated blood from the lungs to the left atrium of the heart.

Pupil The gap surrounded by the iris through which light passes into the eye. Pupil size can be changed by dilation and constriction of the iris.

Putrefying bacteria These break down animal waste and produce ammonia.

Pyramids Diagrams which illustrate quantitatively the relationships between organisms in a food chain. Each organism is represented by a block in the pyramid. Pyramids can show number, biomass or energy relationships.

Reabsorption The way in which substances needed by the body are taken back into the blood from the tubules in the kidney.

Receptors Special cells which are capable of detecting environmental changes.

Recessive The allele which must be present on both chromosomes to show an effect in the phenotype.

Recipient The person who, in a kidney transplant, receives the replacement kidney.

Red blood cells Cells that contain haemoglobin and whose function is to transport oxygen around the body.

Reflex action A rapid automatic response to a stimulus, during which nerve impulses are sent by receptors through the nervous system to effectors.

Glossary

Reflex arc The route taken by a nerve impulse through the nervous system to bring about a reflex action.

Relay neurone (connector neurone) Neurones found in the spinal cord which link sensory and motor neurones.

Renal To do with the kidney.

Renal artery The blood vessel that carries blood to the kidneys.

Renal vein The blood vessel that carries blood away from the kidneys.

Reproduction The formation of offspring.

Resources Natural materials available for the use of organisms.

Respiration The process taking place in living cells transferring energy from food molecules (glucose) to cellular energy.

Respire The cellular process of obtaining energy from food.

Response The reaction of an organism to a stimulus.

Retina The light receptor surface at the back of the eye where light sensitive receptors convert light into nerve impulses.

Rib muscles The muscles between the ribs which contract to raise the rib cage for inhalation.

Root hairs Cells with a large surface area and thin cell wall that absorb water and mineral salts from the soil by osmosis, diffusion and active transport.

Saliva A liquid containing the enzyme amylase produced in the salivary glands.

Salivary glands Glands with tubes emptying saliva into the mouth. Glands in the mouth that secrete saliva and the enzyme amylase.

Sclera The tough outer layer of the eye.

Secondary feathers The flight feathers attched to the bones of a bird's 'hand'.

Selective breeding The process of deliberately breeding animals or plants according to desirable characteristics.

Sensory neurone A nerve cell which carries impulses from sense cells or organs to the spinal cord.

Sexual reproduction This involves two parents who each produce sex cells that must join together. The offspring are genetically different from the parents and each other.

Skeletal tissue This includes bone, cartilage, muscle, ligaments and tendons.

Skin A water-proof, germ-proof layer that contains receptors sensitive to touch, pressure and temperature and plays a part in temperature control.

Small intestine That part of the digestive system where the absorption of soluble foods into the blood occurs.

Soluble Able to be dissolved (usually in water).

Species A group of organisms which look similar and that can breed together to produce fertile offspring.

Spores The reproductive cells produced by fungi.

Sternum The breast bone, which in birds has an enlarged part called the keel, to which the wing muscles are fixed.

Stimulus A change in the environment of an organism which produces a response.

Stomach The part of the digestive system after the gullet where food is churned into a liquid mass.

Stomata (singular: stoma) The tiny openings in the surface of a leaf through which gases can pass by diffusion. The size of the openings is regulated by the guard cells.

Stylets These are the fine, long, usually needle-like tubes found in the mouth parts of most fluid-feeding insects.

Substrate A liquid enzymes can work on.

Suspensory ligaments The muscles in an eye that hold the lens in place.

Swim bladder An air-filled structure used by fish to control their buoyancy.

Synapse The gap between two neurones.

Synovial fluid This is an oily liquid secreted by the synovial membrane.

Synovial joints These are those joints filled with synovial fluid which are adapted to provide as much friction-free movement between bones as possible.

Synovial membranes These secrete the synovial fluid.

T cells These are lymphocytes that contain receptors that destroy antigens.

Target organ The organ affected by the release of a hormone.

Temperature How hot or how cold an object is. Units are °C.

Tendons These attach muscles to bones.

Thorax The chest cavity.

Tissue A group of cells working together to carry out a particular function.

Tissue fluid A liquid formed from the blood plasma and carries soluble substances from the blood to the tissue cells.

Toxic Poisonous.

Toxins Poisons.

Trachea The tube which connects the throat and the lungs and through which air passes into the lungs.

Transpiration The process by which water evaporates from the leaf through the stomata, creating a pull causing water to rise up the plant in the transpiration stream.

Turgor The pressure that the cytoplasm and vacuole of a cell exert on the cell wall.

Urea The breakdown product of amino acids produced in the liver and excreted by the kidneys in urine.

Ureter The tube taking urine from the kidney to the bladder.

Urine The waste fluid produced in the kidneys that contains urea, excess water and salts.

Vaccination (see immunisation) The introduction of small quantities of dead or inactive forms of the pathogen into the body.

Vaccine The dead or inactive forms of the pathogen that are introduced into the body during immunisation.

Vacuole A cavity in the cytoplasm which is surrounded by a membrane. The vacuole contains cell sap.

Variation The differences in characteristics between members of the same species.

Vein A blood vessel taking blood to the heart.

Vena cava The blood vessel that carries blood from the body to the heart.

Ventilation Movement of air in and out of the lungs during breathing.

Ventricles The lower pumping chambers of the heart.

Villi (singular: villus) The finger-like projections in the small intestine that provide a large, thin, moist surface and good blood supply through which the soluble products of digestion are rapidly absorbed.

Viral pathogens A disease-causing virus.

Virus An organism that consists of a protein coat surrounding a few genes.

White blood cells These cells are important in the defence against disease by ingesting bacteria, producing antibodies or producing antitoxins which neutralise the toxins produced by bacteria.

Wilting A condition brought about by loss of water from cells in a plant. The cells cease to be turgid and support for cells and plants is reduced.

Xylem A column of dead cells in a plant that are responsible for the transport of water and mineral ions upwards in the plant.

Index

Note: Glossary entries are in bold

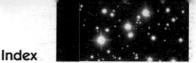